SpringerBriefs in Applied Sciences and Technology

SpringerBriefs present concise summaries of cutting-edge research and practical applications across a wide spectrum of fields. Featuring compact volumes of 50–125 pages, the series covers a range of content from professional to academic.

Typical publications can be:

- A timely report of state-of-the art methods
- An introduction to or a manual for the application of mathematical or computer techniques
- A bridge between new research results, as published in journal articles
- A snapshot of a hot or emerging topic
- An in-depth case study
- A presentation of core concepts that students must understand in order to make independent contributions

SpringerBriefs are characterized by fast, global electronic dissemination, standard publishing contracts, standardized manuscript preparation and formatting guidelines, and expedited production schedules.

On the one hand, **SpringerBriefs in Applied Sciences and Technology** are devoted to the publication of fundamentals and applications within the different classical engineering disciplines as well as in interdisciplinary fields that recently emerged between these areas. On the other hand, as the boundary separating fundamental research and applied technology is more and more dissolving, this series is particularly open to trans-disciplinary topics between fundamental science and engineering.

Indexed by EI-Compendex, SCOPUS and Springerlink.

More information about this series at http://www.springer.com/series/8884

Rüdiger Lohse · Alexander Zhivov

Deep Energy Retrofit Guide for Public Buildings

Business and Financial Models

Energy in Buildings and
Communities Programme

 Springer

Rüdiger Lohse
Linkenheim-Hochstetten
Baden-Württemberg, Germany

Alexander Zhivov
Champaign, IL, USA

ISSN 2191-530X   ISSN 2191-5318   (electronic)
SpringerBriefs in Applied Sciences and Technology
ISBN 978-3-030-14921-5   ISBN 978-3-030-14922-2   (eBook)
https://doi.org/10.1007/978-3-030-14922-2

Library of Congress Control Number: 2019933205

This Springer imprint is published by the registered company Springer Nature Switzerland AG
The registered company address is: Gewerbestrasse 11, 6330 Cham, Switzerland

Preface

The International Energy Agency

The International Energy Agency (IEA) was established in 1974 within the framework of the Organization for Economic Cooperation and Development to implement an international energy programme. The basic aim of the IEA is to foster international cooperation among the 29 IEA participating countries and to increase energy security through energy research, development, and demonstration in the fields of technologies for energy efficiency and renewable energy sources.

The IEA Energy in Buildings and Communities Programme

The IEA coordinates international energy research and development (R&D) activities through a comprehensive portfolio of Technology Collaboration Programmes. The mission of the IEA Energy in Buildings and Communities (IEA EBC) Programme is to develop and facilitate the integration of technologies and processes for energy efficiency and conservation into healthy, low emission, and sustainable buildings and communities through innovation and research. (Until March 2013, the IEA EBC Programme was known as the IEA Energy in Buildings and Community Systems Programme, ECBCS.)

The R&D strategies of the IEA EBC Programme are derived from research drivers, national programmes within IEA countries, and the IEA Future Buildings Forum Think Tank Workshops. These R&D strategies aim to exploit technological opportunities to save energy in the buildings sector, and to remove technical obstacles to market penetration of new energy-efficient technologies. The R&D strategies apply to residential, commercial, and office buildings and community systems, and will impact the building industry in five areas of focus for R&D activities:

- Integrated planning and building design;
- Building energy systems;
- Building envelope;
- Community-scale methods;
- Real building energy use.

The Executive Committee

Overall control of the IEA EBC Programme is maintained by an Executive Committee, which not only monitors existing projects, but also identifies new strategic areas in which collaborative efforts may be beneficial. As the Programme is based on a contract with the IEA, the projects are legally established as Annexes to the IEA EBC Implementing Agreement. At the present time, the following projects have been initiated by the IEA EBC Executive Committee, with completed projects identified by (*):

Annex 1: Load Energy Determination of Buildings (*)
Annex 2: Ekistics and Advanced Community Energy Systems (*)
Annex 3: Energy Conservation in Residential Buildings (*)
Annex 4: Glasgow Commercial Building Monitoring (*)
Annex 5: Air Infiltration and Ventilation Centre
Annex 6: Energy Systems and Design of Communities (*)
Annex 7: Local Government Energy Planning (*)
Annex 8: Inhabitants Behavior with Regard to Ventilation (*)
Annex 9: Minimum Ventilation Rates (*)
Annex 10: Building HVAC System Simulation (*)
Annex 11: Energy Auditing (*)
Annex 12: Windows and Fenestration (*)
Annex 13: Energy Management in Hospitals (*)
Annex 14: Condensation and Energy (*)
Annex 15: Energy Efficiency in Schools (*)
Annex 16: BEMS 1–User Interfaces and System Integration (*)
Annex 17: BEMS 2–Evaluation and Emulation Techniques (*)
Annex 18: Demand Controlled Ventilation Systems (*)
Annex 19: Low Slope Roof Systems (*)
Annex 20: Air Flow Patterns within Buildings (*)
Annex 21: Thermal Modeling (*)
Annex 22: Energy Efficient Communities (*)
Annex 23: Multi-Zone Air Flow Modeling (COMIS) (*)
Annex 24: Heat, Air and Moisture Transfer in Envelopes (*)
Annex 25: Real time HVAC Simulation (*)
Annex 26: Energy Efficient Ventilation of Large Enclosures (*)
Annex 27: Evaluation and Demonstration of Domestic Ventilation Systems (*)

Annex 28: Low Energy Cooling Systems (*)
Annex 29: Daylight in Buildings (*)
Annex 30: Bringing Simulation to Application (*)
Annex 31: Energy-Related Environmental Impact of Buildings (*)
Annex 32: Integral Building Envelope Performance Assessment (*)
Annex 33: Advanced Local Energy Planning (*)
Annex 34: Computer-Aided Evaluation of HVAC System Performance (*)
Annex 35: Design of Energy Efficient Hybrid Ventilation (HYBVENT) (*)
Annex 36: Retrofitting of Educational Buildings (*)
Annex 37: Low Exergy Systems for Heating and Cooling of Buildings (LowEx) (*)
Annex 38: Solar Sustainable Housing (*)
Annex 39: High Performance Insulation Systems (*)
Annex 40: Building Commissioning to Improve Energy Performance (*)
Annex 41: Whole Building Heat, Air and Moisture Response (MOIST-ENG) (*)
Annex 42: The Simulation of Building-Integrated Fuel Cell and Other Cogeneration
          Systems (FC+COGEN-SIM) (*)
Annex 43: Testing and Validation of Building Energy Simulation Tools (*)
Annex 44: Integrating Environmentally Responsive Elements in Buildings (*)
Annex 45: Energy Efficient Electric Lighting for Buildings (*)
Annex 46: Holistic Assessment Tool-kit on Energy Efficient Retrofit Measures for
          Government Buildings (EnERGo) (*)
Annex 47: Cost-Effective Commissioning for Existing and Low Energy Buildings
          (*)
Annex 48: Heat Pumping and Reversible Air-Conditioning (*)
Annex 49: Low Exergy Systems for High Performance Buildings and Communities
          (*)
Annex 50: Prefabricated Systems for Low Energy Renovation of Residential
          Buildings (*)
Annex 51: Energy Efficient Communities (*)
Annex 52: Towards Net Zero Energy Solar Buildings (*)
Annex 53: Total Energy Use in Buildings: Analysis and Evaluation Methods (*)
Annex 54: Integration of Micro-Generation and Related Energy Technologies in
          Buildings (*)
Annex 55: Reliability of Energy Efficient Building Retrofitting–Probability
          Assessment of Performance and Cost (RAP-RETRO) (*)
Annex 56: Cost-Effective Energy and CO_2 Emissions Optimization in Building
          Renovation
Annex 57: Evaluation of Embodied Energy and CO_2 Equivalent Emissions for
          Building Construction (*)
Annex 58: Reliable Building Energy Performance Characterization Based on Full
          Scale Dynamic Measurements (*)
Annex 59: High Temperature Cooling and Low Temperature Heating in Buildings
          (*)
Annex 60: New Generation Computational Tools for Building and Community
          Energy Systems

Annex 61: Business and Technical Concepts for Deep Energy Retrofit of Public
        Buildings
Annex 62: Ventilative Cooling
Annex 63: Implementation of Energy Strategies in Communities
Annex 64: LowEx Communities–Optimized Performance of Energy Supply
        Systems with Exergy Principles
Annex 65: Long Term Performance of Super-Insulating Materials in Building
        Components and Systems
Annex 66: Definition and Simulation of Occupant Behavior in Buildings
Annex 67: Energy Flexible Buildings
Annex 68: Indoor Air Quality Design and Control in Low Energy Residential
        Buildings
Annex 69: Strategy and Practice of Adaptive Thermal Comfort in Low Energy
        Buildings
Annex 70: Energy Epidemiology: Analysis of Real Building Energy Use at Scale
Annex 71: Building Energy Performance Assessment Based on
        In-situ Measurements
Annex 72: Assessing Life Cycle Related Environmental Impacts Caused by
        Buildings
Annex 73: Towards Net Zero Energy Resilient Public Communities
Annex 74: Energy Endeavor
Annex 75: Cost-Effective Building Renovation at District Level Combining Energy
        Efficiency and Renewables

Working Group—Energy Efficiency in Educational Buildings (*)
Working Group—Indicators of Energy Efficiency in Cold Climate Buildings (*)
Working Group—Annex 36 Extension: The Energy Concept Adviser (*)
Working Group—HVAC Energy Calculation Methodologies for Non-residential
Buildings of America.

Additional copies of this report may be obtained from: www.iea-ebc.org;
essu@eia-ebc.org.

Linkenheim-Hochstetten, Germany   Rüdiger Lohse
Champaign, USA   Alexander Zhivov

Acknowledgements

Special thanks are owed to the following individuals, whose invaluable support aided in the publication of this document: William Wolfe, ERDC-ITL, for editing and proofreading; Monika Lohse for proofreading, Jürgen Leuchtner, Annika Burger for graphic design and proofreading.

Contents

Contributing Authors

Dr. Reinhard Jank Ravensburg, Germany

Cyrus Nasseri U.S. Department of Energy, FEMP, Washington, DC, USA

Kinga Porst General Services Administration (GSA), Washington, DC, USA

List of Figures

List of Tables

Chapter 1
Introduction

In European Union countries and in the United States, buildings account for more than 40% of all energy consumed and for 35–45% of CO_2 emissions, making buildings the largest end-use energy sector, followed by industry and transportation [1]. Moreover, over 80% of the existing buildings today are at least 15 years old [2]). Buildings' enormous appetite for electricity —most of which is produced by fossil fuels— threatens our climate, our security, our economy, and our health.

The building sector presents the potential for tremendous improvements in energy efficiency and reductions in carbon emissions [1]. Energy retrofits to the existing building stock represent a significant opportunity in the transition to a low-carbon future. Moreover, investing in highly efficient building materials and systems can replace long-term energy imports, contribute to cost cutting, and create numerous new jobs. Yet, while technologies to improve energy efficiencies are readily available, significant technological progress has not yet been made, and "best practices" for implementing building technologies and renewable energy sources are still relegated to small "niche" applications.

To begin to address this problem, the International Energy Agency (IEA) considers deep energy renovation (DER) and advanced building codes as two top priority goals. To get the building sector on track, it is critical that national policies include three key criteria:

1. a whole-building systems approach with advanced components,
2. the adoption of enforceable building codes, especially in emerging building markets, and application of those codes to component replacement in developed countries, and
3. development of business models to make DER affordable.

In March 2015, the Energy Efficiency Financial Institutions Group ("EEFIG"), tasked by the European Commission and comprised a broad range of stakeholders from the financial services and the energy efficiency community, published a final report [3] that identified the following key issues:

© The Author(s), under exclusive license to Springer Nature Switzerland AG 2019
R. Lohse and A. Zhivov, *Deep Energy Retrofit Guide for Public Buildings*,
SpringerBriefs in Applied Sciences and Technology,
https://doi.org/10.1007/978-3-030-14922-2_1

Table 1.1 Major barriers to DER

Level	Barriers to DER
Market	Market organization–price distortions prevent building owners from appraising value of energy efficiency (EE) measures Split incentives: investor cannot capture benefits of DER investments Changing policy and market conditions do not enable confidence in long term investment decisions
Financing	Upfront costs and dispersed benefits. DER is complicated, risky, and has high transaction costs Lack of awareness of potential financing entities
Information	Lack of sufficient information such as Energy Performance Certificates to prepare rational investment decisions
Regulatory/institutional	Low energy prices that discourage energy efficiency investments Institutional bias toward supply-side instead demand-side investments Lack of sufficient business models with incentives for DER and life cycle costs Lack of sufficient long term strategies to deploy DER in building stock Perception of risks allocated in DER investment programs due to the uncertainties of predicted energy cost savings and the lack of evaluated projects and default analysis Lack of standardized protocols for de-risking. Standardized evaluation methods for measuring and verification is still lacking
Technical	Insufficient capacity to develop, implement, maintain high efficient Energy Conservation Measure (ECM) bundles

- DER of buildings provides many benefits that should be fully captured and concretely communicated with real-world examples to the most important financial decision makers, such as public authorities, real estate owners, and managers.
- Data is a key stumbling block. Processes must be established in a way to ensure that data are reliable and transparent. Information must be provided in a way that links potential real estate value increases with respective efficiency investments. A key element is to standardize data and processes to achieve greater investment in energy efficiency.

Table 1.1 summarizes major barriers for implementation of DER projects as identified in [4, 5].

Research under the IEA EBC Program Annex 61 has been conducted with a goal of providing a framework, selected tools, and guidelines to significantly reduce energy use (by more than 50%) in government and public buildings constructed before the 1980s with low internal loads (e.g., office buildings, dormitories, barracks, public housing and educational buildings) undergoing major renovation.

Best practices from Europe (Austria, Denmark, Estonia, Germany, Ireland, Latvia, Montenegro, The Netherlands, United Kingdom) and the United States have been studied and 26 examples of implemented retrofit projects, in which site energy use has been reduced by 50% or better compared to pre-renovation base line, have been documented in the "Deep Energy Retrofit—Case Studies" report [6]. These case studies were analyzed with respect to energy use before and after renovation, reasons for undertaking the renovation, co-benefits achieved, resulting cost effectiveness, and the business models followed.

A list of core energy efficiency technologies was generated from the results of case studies, from surveys and discussions conducted at the American Society of Heating, Refrigerating, and Air-Conditioning Engineers (ASHRAE) Technical Committee (TC) 7.6 "Public Buildings" working group meetings in 2013 and 2014, and from previous experience and research conducted by the Annex 61 team members. These technologies, when applied together (as a bundle), will reduce the total building site energy use by about 50% (including plug loads). Technical characteristics of these building envelope-related technologies grouped into a "core technologies bundle" have been studied through modeling and life cycle cost (LCC) analysis for representative national climate conditions and presented in the "Deep Energy Retrofit—A Guide to Achieving Significant Energy Use Reduction with Major Renovation Projects (DER Technical Guide)" [7]. Results of these studies provided a base for setting minimum requirements to the building envelope-related technologies to make Deep Energy Retrofit feasible and, in many situations, cost effective. Other characteristics of these technology bundles are based on the requirements of national standards or on best international practices, which have been collected and summarized and presented in the DER Technical Guide.

Use of energy efficiency measures (EEMs) in combination with "bundles" of core technologies and high-efficiency appliances will foster further reductions in energy use. The DER Technical Guide also provides examples of "best practices" that illustrate optimal methods of applying these technologies in different construction situations.

Since the funding available for public buildings refurbishment is limited, a major obstacle to the implementation of energy retrofit projects beyond minimum requirements is a lack understanding of how to make DERs cost effective. Implementation of DER projects is also limited by the organizational and technological capacities of the public building agencies. To address these needs, Subtask B of Annex 61 has focused on collecting and analyzing the information related to DER project organization, project financing, reduction of investment costs, and on the monetary quantification of non-operational benefits resulting from these projects

The "Deep Energy Retrofit Business Guide (DER Business Guide)" resulted from the Subtask B describes the facilitation and implementation of DER projects from the business perspective. The evaluation of DER case studies indicates that cost-effectiveness and availability of funding are the most relevant decision-making criteria to initiate a DER concept. The implantation of cost-ineffective DER concepts is one major problem that often leads to shallow refurbishments or staged refurbishment projects. The key performance indicators of the cost effectiveness in the building and real estate sector are the cash flow analysis and the net present value (NPV).

Thus the DER Business Guide considers strategies to improve the cost effectiveness of DER projects by reducing the cost of investment and by quantifying energy- and non-energy-related cost savings. The first approach to reduce investment costs is to consider available grant programs. The next step is to devise least-cost planning approaches that cost-effectively combine the investment costs of energy conservation measures (ECMs) in DER core technology bundles.

However, especially in countries with low energy prices, it can be difficult to achieve cost effective buildings through energy savings alone. To address such situations, the DER Business Guide presents methods of using fuel switching and the integration of renewable energies to not only improve the cost effectiveness but also increase the energy resilience of buildings and neighborhoods by decreasing their dependency on the grid power supply. In addition, the DER Business Guide presents major non-energy-related operative cost savings: avoided maintenance costs for replaced worn-out equipment, reduced operation costs after the implementation of building automation services, and reduced building and machinery insurance risk premiums. On the revenue side, DER augments the monetary value of building properties by increasing the market value of the property (which can justify higher rental rates and sales prices), and by increasing available and usable net floor space. Best practice examples show that a DER can increase energy cost savings by 30–120%, dramatically improving the cost-effectiveness of the DER. Recent projects that quantified the impact of DER on the indoor climate and on the occupant satisfaction in monetary terms, determined that a DER may add another 50–150% to the value of energy cost savings.

Another major hurdle for the implementation of DER that the DER Business Guide addresses is the lack of available public funding. Often credit lines are limited, equity funding is scarce, and the DER must compete with many other public projects. To increase the availability of equity and private sector funding, the DER Business Guide shows how carefully implemented quality assurance (QA) can help to improve DER projects so they are seen as sound, reliable investments.

So far the majority of DER projects have only been implemented in a traditionally funded and conducted business model, which is limited to the financial, organizational, and technological capacities of public building agencies. The Business Model Guide provides information on advanced energy performance contracting (EPC) business models that combine financing, implementation, and operation services with performance-related remuneration. To date, EPC has not been used in DER projects; however, the Guide highlights recent efforts to advance the EPC mechanism and broaden its scope to include ambitious DER projects.

The "DER Business Guide" includes examples of pilot projects implemented by innovators in Belgium, the United States, Latvia, and Germany.

The target audience for this Guide includes energy managers, project managers from Energy Service Companies (ESCOs), general contractors, and financing organizations involved in funding energy projects. It will also be of interest to building owners, executive decision makers from public, Federal, State and local governments, military administrators, architectural and engineering firms, and manufacturers of energy efficient products and systems.

Table 1.2 Historical improvement of the ASHRAE Standard 90.1

ASHRAE standard 90.1 version	1975	1980	1989	1999	2001	2004	2007	2010	2013
Energy use index	100	100	86	81.5	82	69.7	65.2	46.7	43.4

Table 1.3 Historical improvement in European national energy requirements for buildings without plug-loads

Country	Pre-1980 Energy use intensity (EUI)	Current national standard	Energy intensity use of current national standards
Denmark	Dwellings: 167.1 kWh/m^{2}year [9]	BR10 [8]	Dwellings: 52.5 kWh/m^{2}year +1650 kWh/GFA Office: 71.3 kWh/m^{2}year + 1650 kWh/GFA
Germany	WSVO 1977 [11] Dwellings 150–250 kWh/m^{2}year [12] Schools: 210 kWh/m^{2}year [10]	EnEV 2014	Current: Energy Ordinance (EnEV 2014 for new buildings Refurbishment: + < 40%) Dwellings (new): 50–60 kWh/m^{2}year Schools new/refurbished: 80–125 kWh/m^{2}year [11]
Austria	Maximum U-values	OIB RL 6 [16]	Heating residential buildings: max. 87.5 kWh/m^{2}year; Nonresidential buildings: max. 30 kWh/m^{2}year

What is Deep Energy Retrofit? Though the Deep Energy Retrofit erm is widely used, there is no established global definition. Since the energy crisis of the 1970s, energy requirements pertaining to new construction and building renovation worldwide have significantly improved. Tables 1.2 and 1.3 list standards and requirements used to design and construct buildings over the past ~50 years (pre-1980s to today). Since the 1980s, building energy use requirements in the United States (Table 1.1) have improved by more than 50% (calculated without consideration of plug loads). Further, buildings and building systems degrade over time with cracks in the building envelope; dirty and leaky ducts; lower efficiencies when heating, ventilating, and air-conditioning (HVAC) systems are not regularly commissioned, etc. This can reduce their energy performance by at least 10%. It is technically feasible to recoup these inefficiencies and further reduce building energy use by more than 50% using technologies readily available on the market by simply adapting current requirements for new buildings to the refurbishment of the existing building stock.

Analysis conducted by the IEA EBC Annex 61 team shows that a significant number of commercial and public buildings have reduced their energy consumption by more than 50% after renovation, and that some have met Passive House Institute energy efficiency standard or the net zero energy state. According to the Global Building Performance Network prognosis [13], a DER that follows the most recent and proposed EU guidance can improve the buildings energy performance by at least 80%.

Based on these experiences, the IEA EBC Annex 61 team has proposed the following definition of the Deep Energy Retrofit:

Deep Energy Retrofit (DER) is a major building renovation project in which site energy use intensity (including plug loads) has been reduced by at least 50% from the pre-renovation baseline with a corresponding improvement in indoor environmental quality and comfort.

A DER requires a whole-building analysis approach along with an integrative design process. A "whole-building analysis" means that the building is considered as a single, integrated system rather than as a collection of standalone systems, such as building envelope, HVAC system, renewable energy system, building operations, etc. The whole-building approach facilitates the identification of synergistic relationships between the component systems. Analyzing systems in isolation does not effectively identify synergies between systems. For example, improving the building envelope, providing solar heat gain control, and improving lighting systems could substantially reduce a building's heating and cooling energy demand. This would in turn reduce the required size of duct systems, air-handling units, boilers, and chillers. Likewise, replacing an aging air-handling unit with a smaller, more efficient unit could improve indoor air quality and further reduce energy demand. Such cascading benefits would not be achievable if the building were not analyzed as an integrated whole.

The key to whole-building analysis is the use of an integrated design process. The whole-building analysis differs from a traditional design process in that it brings all relevant disciplines together for an initial charrette-based study of the problem as a whole, based on collaboration and shared information, whereas a more traditional process is based on a linear flow of information passing from one discipline to another.

1.1   Major Renovation and Deep Energy Retrofit

The need to reduce energy consumption is one of many reasons buildings undergo major renovations. Some of the most common reasons are:

- Extend the useful life of a building with an overhaul of its structure, internal partitions, and systems
- Repurposing of the building (e.g., renovation of old warehouses into luxurious apartments)
- Bringing the building to new or updated codes such as fire protection

- Remediation of environmental problems (mold and mildew), improvement of the visual or thermal comfort, or indoor air quality
- Adding to the value with improvements to increase investment (increasing useful space and/or space attractiveness/quality) resulting in a higher sale or lease price.

The U.S. Department of Energy (USDOE) (2010) and EPBD (2016) define a major building renovation as any renovation the of which cost exceeds 25% of the replacement value of the building. EPBD must be considered in renovations with more than 25% of the surface of building envelope undergoes renovation.

Timing a DER to coincide with a major renovation is best since:

- during the renovation,
- the building is typically evacuated and gutted;
- scaffolding is installed;
- single-pane and damaged windows are scheduled for replacement;
- building envelope insulation is replaced and/or upgraded;
- and most of mechanical, electrical lighting, and energy conversion systems (e.g., boiler and chillers) along with connecting ducts, pipes, and wires are replaced.

A significant sum of money covering the cost of energy-related scope of the renovation designed to meet minimum energy code (a significant part of the DER) is already budgeted anyway.

1.2 Deep Versus Shallow Energy Retrofit

Current studies show that the typical approach to refurbish the building stock is to follow a "shallow renovation" track that focuses on single measures, partial refurbishments, primarily on lighting retrofits, HVAC replacement and retrocommissioning,, and other ECMs that provide low risk and short payback periods. Such projects rarely include measures such as façade and roof insulation, replacement of windows, remediation of thermal bridges, or significant improvements in building air tightness. In countries with stringent source energy targets for refurbishment projects, building owners tend to choose renewable energy and heating supply solutions over measures that improve energy efficiency, increase insulation, etc.

From the perspective of a public building owner, shallow refurbishments, especially HVAC replacements, offer a large risk for a "missed opportunity" if envelope improvements such as façade upgrade, or roof or window replacement are not undertaken. A combined approach would have allowed a downsizing of the HVAC system due to lower heating and cooling demands, and elimination of perimeter zone conditioning, and would likely have provided improved comfort. More importantly, the findings from the Annex 61 pilot case studies show that a combined bundle of HVAC, thermal envelope, and renewable power and heat supply with individual short- and long-term payback periods are likely to be cost effective. Moreover, picking the most cost effective ("cream-skimming") measures such as HVAC and other short-term options will make future investments for remaining items even less appealing since the shortest term investment would have already ready been done [14]. For the decision-making on the building level, it is necessary to identify cost-effective pathways for DER instead of considering minimum requirements in "shallow refurbishment" approaches.

This is also true on the macroeconomic level. Dynamic simulations [15] have clearly shown that "deep renovation as a carefully phased process " is a more promising strategy to reach long term (2050) climate targets than "shallow renovation at high speed." "Shallow renovation" with very high shares of renewable energy undertaken to achieve source energy targets appears to be 3.5% more expensive.

DER Core Technologies Bundles

Previous research conducted under the IEA EBC Annex 46 identified and analyzed more than 400 individual EEMs that can be used when buildings are retrofitted. Measures include those related to the building envelope, mechanical and lighting systems, energy generation and distribution, internal processes, etc. [16].

However, the implementation of some individual measures over time is not the most cost effective path to achieving energy savings of 50%. The common practice, which dictates choosing the most cost-efficient measures first (using the strategy of "single measures") is counterproductive in that it reduces the cost-effectiveness of a holistic refurbishment approach.

To achieve a DER, it is necessary to implement a bundle of individual measures (such as building envelope insulation, window replacement, improved airtightness, etc.). This comes at a price of increasing investment costs and payback periods. The DER requires considerable efforts to detect the cost-benefit optimum of the energy target value and the investment of the measure bundle. IEA EBC Annex 61 Subtask A assessed ECM bundles from DER projects that provide pathways toward cost-effective measure bundles to optimize the cost-benefit ratio.

Table 1.4 provides a list of core energy efficiency technologies was generated from the results of case studies of DERs conducted in Europe and North America. [17] These technologies, when applied together (as a bundle), will reduce the total building site energy use by about 50% site energy including heating and power. Technical characteristics of these building envelope-related technologies grouped into a "core technologies bundle" have been studied through modeling and LCC analysis for representative national climate conditions.

Characteristics of some of these core technology measures depend on the technologies available on an individual nation's market, on the minimum requirements of national standards, and on economics (as determined by an LCC analysis). In most cases, a DER feasibility study will assess benefits resulting from the reduction of energy load demand and consumption. In countries with low energy prices, additional measures will be necessary to increase the cost effectiveness for building owners and investors: the inclusion of renewables can improve energy security and the overall cost-effectiveness of the DER approach.

Table 1.4 Core technologies bundles for DER

Category	Name
Building envelope	Roof insulation
	Wall insulation
	Slab insulation
	Windows
	Doors
	Thermal bridges remediation
	Airtightness
	Water vapor barrier
	Building envelope quality assurance protocols
Lighting and electrical systems	Lighting retrofit daylight, zoning and presence control systems
HVAC	High performance motors, fans, furnaces, chillers, boilers, etc.
	Dedicated outdoor air system (DOAS)
	Heat recovery (sensible and latent)
	Duct insulation
	Duct airtightness
	Pipe insulation

References

1. World Energy Outlook 2013, IEA, Paris 2013. http://www.worldenergyoutlook.org/
2. IEA, Technology Roadmap Energy Efficient Building Envelopes, 2012. https://www.iea.org/publications/freepublications/publication/TechnologyRoadmapEnergyEfficientBuildingEnvelopes.pdf
3. Energy Efficiency the First Fuel for the EU Economy, EEFIG, Brussels 2015. http://www.unepfi.org/fileadmin/documents/EnergyEfficiency-Buildings_Industry_SMEs.pdf
4. Energy Efficiency Finance Group (EEFIG), Final Report, Brussels, 2015
5. Investors Day 2014, KEA, Karlsruhe, 2014
6. DER Case Study Report, IEA Annex 61.org
7. DER Technical Guide, IEA Annex 61.org
8. http://www.buildup.eu/sites/default/files/content/BR10_ENGLISH.pdf
9. DER Case Study Report, IEA Annex 61.org
10. http://www.bbsr-energieeinsparung.de/EnEVPortal/DE/Archiv/WaermeschutzV/WaermeschutzV1977/Download/WaermeschutzV77.pdf
11. R., Lohse. EDLIG Report Arbeitspaket B, Karlsruhe, 2014: Some example buildings show that much lower number are possible for school and office buildings as well (< 40 kWh/m^{2}y), see further German reports within Annex 61
12. http://www.gbpn.org/reports/delivering-energy-savings-buildings
13. R., Lohse. EDLIG Report Arbeitspaket B, Karlsruhe, 2014 Some example buildings show that much lower number are possible for school and office buildings as well (< 40 kWh/m^{2}y), see further German reports within Annex 61
14. http://www.gbpn.org/reports/delivering-energy-savings-buildings

15. K. Bettgenhäuser, Rolf de Vos, J. Grözinger T. Boermans. 2014. Deep renovation of build-
    ings An effective way to decrease Europe's energy import dependency. Ecofys project #
    BUIDE14901by order of Eurima; Levin Nock and Clint Wheelock. 2010. Energy Efficiency
    Retrofits for Commercial and Public Buildings.; b) Energy Savings Potential, Retrofit Business
    Cases, Financing Structures, Policy and Regulatory Factors – Demand Drivers by Segment,
    and Market Forecasts. Pike Research. C) Andreas H. Hermelink and Astrid Müller. 2011. Eco-
    nomics of Deep Energy Renovation. Implications of a set of Case Studies. Ecofys project #
    PDEMDE101646 by order of EurimaA, Capturing the Multiple Benefits of Energy Efficiency,
    2014
16. American Society of Heating, Refrigerating, and Air-Conditioning Engineers (ASHRAE) Stan-
    dard 100 (ASHRAE 2015)
17. DER Case Studies, Moerck et al. 2017, IEA presented at; ASHRAE Technical Committee (TC)
    7.6 "Public Buildings" working group meetings in 2015

Chapter 2
The Scope of DER and Investment Costs

2.1 Development of the DER Scope from the Business Perspective

As stated in Sect. 1.2, DER is usually conducted as a part of the major renovation project. Most major renovation projects include a scope of work, which can be either non-energy-related or energy-related. A non-energy-related scope of work may include elements such as different construction jobs related to changing floor layouts (e.g., moving/removing internal partitions), adding bathrooms, removing asbestos, and adding sprinkler system.

An energy-related scope of work of a major renovation project typically includes replacement of existing mechanical, lighting and electrical systems, replacement of some or all windows, replacement of existing ductwork and plumbing systems, etc.

The project starts with a formulation of detailed definition for project requirements and criteria such as the Statement of Work (SOW) or Owner's Project Requirements Document. These documents establish the basis for the design of the building, against which tenders (i.e., bids) will be made for both design and construction services. Acceptance of these criteria indicates verification of understanding of the criteria by those proposing to provide design or construction services (see Appendix H of the Annex 61 Technical Guide).

There are currently standards and guidelines in place through many standards and governmental organizations that provide detailed guidance for development of these documents.

The development of the energy-related SOW starts with the data gathering phase, which includes all necessary information about the building or a cluster of buildings planned for a major renovation project.

R. Lohse and A. Zhivov, *Deep Energy Retrofit Guide for Public Buildings*,
SpringerBriefs in Applied Sciences and Technology,
https://doi.org/10.1007/978-3-030-14922-2_2

Engineering grade energy audit shall be conducted to obtain energy data, required to develop the project baseline, and to evaluate conditions of the building envelope, mechanical, electrical, water supply and sewer systems, along with conditions of energy generation equipment (e.g., boilers and chillers with decentralized energy systems), or types and parameters of energy utilities when the building is connected to district systems. The composition and requirements to building audits is determined in ISO 50002 (2014), DIN EN 16247 (EU/Germany), AS/NZS 3598:2014 (Australia) and Procedures for Commercial Building Energy Audits (USA) [1].

Documented data of the building energy use obtained from energy metering and submetering can be used as a baseline to calculate EUI and to gain a better understanding of how energy is used and wasted in the building by different systems. The baseline is typically used to compare building energy use after and before its renovation, but it is not used for making a business case for the DER project. A **major renovation** with the energy-related scope of work is usually required to **meet current minimum standard** requirements and is considered to be a **base case** for the life-cycle cost analysis of renovation scenarios and their cost optimization.

Competing DER project scenarios may differ in the level of energy use reduction to be achieved compared to the base case. **Energy-related investment costs will usually be higher in a DER** scenarios compared to the base case. Some energy-related improvements included into DER, e.g., building envelope insulation and mitigation of thermal bridges, installation of high performance windows, and air tightening the building envelope, are expensive and are rarely included in the scope of major renovation or are performed following lower energy performance requirements. However, reduction of heating, cooling or humidity loads resulting from implementation of these measures will result in the need for a **smaller and sometimes simpler HVAC system, which will, in turn, reduce both initial investment and capital replacement costs related to these systems.**

Energy use reduction that can be achieved with different scenarios can be evaluated using computer-based energy simulation models. Modeling results must be calibrated if there is a difference in the building use after renovation, if there are new requirements to thermal conditions, or if there is a change in the plug loads. Energy modeling also allows to optimize building specific technical characteristics of technologies to be used in the DER project such as U-values of walls, windows, roofs (to be compared to those recommended in the Technical Guide [2]). Information obtained from the scope of the selected scenario and characteristics of technologies to be used in the project, can be used to drive investment costs.

Pre-renovation Building Information

Technically sound energy utility data provides the critical starting point for accurate projection of potential energy savings as well as for measurement after retrofits and retrocommissioning. It is also critical for determination of proper goals, and for compliance with these energy goals. At a minimum, the following pre-renovation building data are necessary:

	Type of information	Description/Example
1	Building usage data	Usage and operation schedule before renovation, number of occupants, occupant schedule, and activity level before the renovation
2	Life cycle cost baseline	For example, energy costs, operation costs, maintenance costs, secondary investment costs
3	Global and specific end energy usage	In order to depict the global and specific consumption the sub-metering structure must be consistent and the data have to be collected
4	Adjusted end energy usage	For the baseline period defined, the global and specific consumption is adjusted at the hand of annual and long term heating and cooling degree days
5	Baseline of operational/performance data:	System performance data collected through interviews, review of building documentation, spot measurements
6	Load profiles	In order to receive information on the energy load demand and saving potentials the load profiles of energy supply and major energy consumers need to be evaluated
7	Indoor building climate	Collection of requirements to indoor air temperature, ventilation rate, and relative humidity before the building renovation
8	Design and as-built-drawings	Showing construction of existing walls, floors, and roofs Building orientation, building elevations showing existing opening/fenestration characteristics, including materials, anticipated performance characteristics, sizes, numbers, and locations
9	HVAC systems	Inventory of currently operated HVAC systems with some major operation parameters such as electric load, temperatures, etc.
10	Plug loads	For the evaluation of plug load energy consumption the types, the quantity, and power consumption requirements of major plug loads are collected
11	Building area breakdown	Building area breakdown by end-use, HVAC sequences of operation, Interval submetering of separate equipment or processes, at least 1 week
12	Simulation calculation	Recalibration of a pre-renovation building energy model against utility data

For more information, see Appendix H of the Technical Guide [2].

2.2 Determination of Technical Concept and Investment Costs

One of the major risks in determining the cost effectiveness of the DER is the relia-bility of the modeling results (scenario and savings) and of the investment costs. At the end of this step, the ECM bundles and energy supply will be determined from a modeling process according to the requirements of the decision-making criteria. The technical result will be combinations of ECM and supply measures for each of the building scenarios. Using specific investment cost data, the investment costs can be calculated and the investment cost optimization can be prepared. From the energy balance of the modeling process, the energy costs in the post-refurbishment phase will be available. From the comparison with the baseline energy costs, the energy savings will be calculated. QA is required to de-risk the investment cost estimates and the quality of the modeling predictions for the business model.

Definition of Building Level Targets

The setup of a decision-making matrix will have a strong impact on the investment costs, the LCCs, the financial instruments and the business model. At the end of this step, users will have access to a matrix of decision-making criteria derived from an overall strategy, e.g., an energy master plan that allows the building owner to conduct and assess feasibility studies and prepare the major investment decisions in a DER project.

   Future usage plan: The public sector with its continuously changing social, economic framework conditions must consider the future usage of the buildings; which buildings will still be in operation in the future and what will be the purpose of the building? The predictions will be related to results of demographic modeling processes and to commercial aspects such as market value and environmental criteria.

Energy and overall targets of a DER project:

- To steer the complex process of a DER project development, a decision-making matrix will be developed that will be used to prepare the decision-making for a technical concept, and will also be used during the major steps of planning, design and implementation of the DER project. In a DER project, the decision-making of DER projects balances technical, environmental, and economic metrics.
- **Economic metric**: The scope of costs (e.g., investment) and revenues (e.g., site energy savings, maintenance) will be evaluated over a defined time-frame by the NPV: all expenditures and revenues of the DER project are cumulated over the payback term on today's value.
  The timeframe is restricted by the technical life cycle period for the most valuable components of the project, the interest rate conditions over time, the requirements of an investor, or other limiting factors. In the United States, Federal Government payback terms may be up to 25 years. In the German Federal Government and also in the majority of the German municipal sector, refurbishments are limited to payback periods of 20 years.

- **Site- or source-energy target**: In many countries with a Near Zero Energy targets, national building regulation have defined source energy targets. The major decision making criterion in the refurbishment process, however, is the cost effectiveness which reflects the impact of a DER on the end energy target. A refurbishment project with only a few demand-side measures can achieve excellent source energy balances by implementing renewables and cutting back the power input from the regional grid; however, the site energy impact may be not sufficient for national source energy requirements. In order to reflect both site and source energy target values (EUI in kWh/m^2yr) have to be defined at the beginning of the simulation process.

- **Carbon-footprint**: The public sector should take the role model in terms of reducing the carbon footprint in the building stock; the decision-making matrix may assess the carbon footprint of the energy demand reduction. The overarching weighting of contradictory targets of economic, environmental and technologic targets is required to compare and select from different DER concepts and compare the best concepts of different buildings. Table 2.1 shows a decision-making matrix of a recently accomplished DER in Germany. After each major step of the DER project development and implementation, the matrix was used to control the performance of the project so far.

Develop technical understanding by building simulation:

The modeling will be carried out to find out in which way the DER targets can be implemented in a technical way in two stages: (1) definition of performance indicators to achieve the energy target values and (2) selection of adequate measures and allocate specific investment costs for each of the DER measures.

First stage is to define the system boundaries of the modeling: the system boundaries include the building with its U-values, heating, and electricity demands (I) and the energy distribution and conversion (II). The modeling process (Table 2.2) includes the "baseline," the "base case" with the national minimum requirements for building refurbishments; in addition a DER scenario and an "ambitious scenario" such as "passive house" will be defined.

For the investment cost estimate a functional specification of the major components (wall, roof, basement insulation, windows, lighting, ventilation, and supply options) must be defined. The non-energy-related investment costs, i.e., repurposing costs and additional benefits such as increasing usable floor space, etc. will be collected for the investment cost estimate. QA measures that have been defined in recent years [3] allow a crosscheck of methods, inputs, and outcomes of the modeling results from the investor's perspective (i.e., without diving deep into the calculations).

The complexity of simulation models does not allow more than a short description in this context. However, the modeling approach must be selected carefully by the sort and precision of results it is expected to provide at the end of the process, by available data (Table 2.3).

Table 2.1 Example of a DER target and decision-making metrics for case study building

Crit. Metric criteria	Description	Key Performance Indicators (KPI)	Weighting factor
1	Life-Cycle Analysis (LCA) assessment (life time period of 20 years, average capital interest rate 3.5%) → Goal: economically optimized combination of energy conservation and supply, low LCA approach. Includes: capital and operation costs, revenues from energy, maintenance and operation cost savings	NPV of cost and revenues over 20 yrs	65%
2	Technical quality of concept and measures (endurance of technical measures [40%], reduction of source energy 60%) → Overall target: sustainable investment with high residual value and excellent source energy balance	Evaluation points between 0 and 100	15%
3	Carbon footprint (assessment of carbon emission from energy consumption and production with given factors/MWh 100%) → Overall target: increase efficiency, use of renewables	CO_2 (t/yr)	10%
4	Indoor climate and condition (assessment of noise reduction effect [40%], of air exchange rate and air quality concept [40%], cooling concept [20%]) →Overall target: increase indoor climate, collaboration with building users	Credit points	10%

Select a modeling tool with regard to accuracy of results:

Models in use are dynamic thermal simulation models (EnergyPlus, TRNSYS) as well as steady-state or quasi-steady state models mostly in use in EU (PHPP, DIN 18599 Tools). Compared to dynamic modeling tools, the quasi-steady state tools are based on Excel™ spreadsheets, which are accessible to most energy consultants without special skills required by more sophisticated software, e.g., TRNSYS.

These monthly based modeling tools should be able to consider several usage zones, the impacts of combined and interfering ECMs such as HVAC systems, the thermal envelope, low-energy heating distribution, energy supply systems, and processes with internal loads and lows. In addition, specific modeling tools

Table 2.2 Modeling scenarios in case study office building

Scenario	Energy savings/target site EUI	Scope of investment	QA measures
0 "baseline"	Current performance	Reinvestment to keep building operable, remove major technical issues, no DR	Compare modeling input and output to results from step 2 (baselining): energy baseline, cost baseline, audit
1 "base-case"	Building code for building stock: 100–120 kWh/m^2 heating	This is the DR scenario: technical measures non-energy-related and minimum energy requirements	U-values and target EUI on input and output of modeling
2 "new building" (equates the DER −50% scenario here)	Adoption of building code for new buildings to refurbishment 50–60 kWh/m^2 heating	Delta-investment in comparison to "1" for advanced U-values and performance parameters of heat recovery, lighting	Check plausibility of U-values, target EUIs on input and output of modeling
3 "passive house"	Adoption of PH standards 30–45 kWh/m^2 heating	EUI, thermal bridges and air-tightness, additional QA	

Table 2.3 Criteria to select reliable energy modeling tools [4]

Criteria	Sub-Criteria
Model applicability	Flexibility and robustness. Framework definitions of the case study, recalibration possible, User friendliness.
Approach selection	Does the model fit to the purpose of the case study? Does it depict environmental standards required in the study? How flexible and adaptable is the model?
Quality and accuracy	Assumptions and methodology of the model fit the case? Which stages are foreseen in the model and which quality of data is delivered? Which methods exist to validate results and procedures?
Data availability	Is our data inventory sufficient for the model?
Desired outcomes	Parameter sensitivity analysis and key design messages the model should be able to provide.
Staff	Is the modeling conducted by an experienced expert; is recalibration data available and is the expert able to use those and document the calculation steps.

exist to advance results for single technical applications such as lighting (ReLux, LightSpace), ventilation systems, combined heat and power (CHP), biomass, solar heating, and photovoltaic (PV) (BHKW-Plan, TSOL).

The results however of dynamic models are likely to be more accurate: the simulation results of passive houses compared with the post-refurbishment performance shows that dynamic hourly modeling may provide more accurate data that are closer in comparison of the simulation tools working on a monthly basis (practical experience of simulations in EBC Annex 61 shows a better accuracy of 5-10% for hourly data based simulations). However, the comparison did only consider low-medium equipped office and housing buildings. For these building types, the sensitivity analysis shows, however, that this level of accuracy is, in most cases, sufficient for the design of a business case.

QA in the modeling process:

Modeling processes are work intense, complex and must be carried out and documented appropriately with regard to the QA and the reuse of data sets in the detailed design process.

The evaluation of accomplished DER projects show that the modeling results and the building performance do not often match. To achieve a high degree of reliability, the saving calculations have to follow a vetting process with the following approaches:

- Crosscheck modeling results with target EUIs and U-values from recalibrated modeling projects.
- Calculations of estimated savings for projects of the scale anticipated must be based on "open-book" calculation methods or tools and will be used to perform verified savings calculations as part of the measurement and verification (M&V) effort, using post-retrofit monitored data.
- Energy conservation measure descriptions: descriptions of the existing conditions, proposed retrofit, and potential interactive effects for each measure under consideration.
- Recalibration: The pre-retrofit energy consumption estimated for each system involved in an ECM must be compared to the estimated or measured end-use energy usage to ensure that the estimated energy usage is in line with baseline estimates. Similarly, estimated energy savings should be compared to empirical data such as previous energy savings estimates or M&V data from accomplished DER projects for reasonableness.
- Calculation Process Description: All significant steps in the calculation should be sufficiently described so that a reviewer can reconstruct and assess the calculations in the decision-making phase of a DER project.
- Economic benchmark analysis of ECMs: Ascertain and record the return on investment (ROI) criteria best expressed for simplicity as a NPV of a cash flow analysis of each ECM measure bundle individually.
- Consider synergetic effects: when ECM bundles are modeled, the results have to be revised with regard to the synergetic effects between ECMs and energy supply measures.

- The transformation of results to other buildings must consider that the results of modeling will always be related to the individual investment and LCA cost and demand structures, the data inventory, and the use of the building.
- Final quality check of results: the calculation outcomes (investment costs (next chapter), end energy usage savings) should be compared to similar projects for reasonableness.
- With regard to the reliability of the energy-saving calculation, the "rebound effect" must be considered carefully. This effect describes the fact that modeling of the existing building, in most cases, leads to a theoretical demand that has a variation of 20–30% (in the case study 16%) compared with the measured energy consumption provided in the utility bill. This is specifically the case for buildings built before the 1960s and if a modeling does not allow depicting different indoor temperatures over a time period of 24 h over a reference year. The same effect can be noticed also for the results after refurbishment. On the other hand, buildings built in the last 20–30 years tend in the modeling process to the reverse effect (prebound). The ultimate cause [5] is the variation between modeled operation and utilization parameters from the actual performance. In old buildings, the average room temperature is kept lower than in modernized buildings. Also the data inventories of U-values, construction materials, thermal bridges etc. is mostly not adequate enough to set up the modeling on a reliable basis. To reduce this effect rebound factors will have to be used in modeling approaches.

2.3 Determination of DER Investment Costs

To prepare the assessment of the economic validity of the different combinations of ECM and supply measures, the investment costs must be identified and analyzed. For decision-making it is important to present the investment and annual cost at the first place. To compare different energy efficiency scenarios is more likely to focus on the "delta-cost," which is the incremental investment and annual costs for a better energy standard than the basic requirements.

Investment costs: The investment costs will consider the primary and the secondary investments. The investment costs are estimated using functional specifications derived from the modeling approach. This is to be understood as the association of installation **costs** to each technology, including material and labor costs, removal and disposal of existing technologies, business profit and general expenditure. Embedded or grey energy is not considered.

Primary investment costs are the initial investment costs to set up the ECM and supply measures. By using net present value factors, considering the timeframe (see below) and a constant capital interest rate over time the investment costs are transformed into constant perennial capital costs (annuities).

Secondary investment Costs: are the investment costs for maintenance and refurbishment (M&R) of the building and its components. Usually DER projects will have a payback period > 20 years. Such long time periods exceed the technical life of some moving parts like pumps, electric valves, and control systems. These may have to be replaced in part or in whole. Secondary investments consider these spare parts or complete replacement in the timeframe. The secondary investments, which will take place in the future, will have to consider increasing investment costs, which are usually adopted by investment cost indexes on national level.

Example: An investment over 12 years is then the value of the investment today multiplied by the cumulated investment cost index from year 1 to 12.

$$C \sec(tn = 12\,a) = \sum Index(yr\ 1 - 12)^* C \sec(2015) \tag{2.1}$$

Gathering of accurate investment costs

Two risks must be considered in terms of accuracy of investment cost calculation: (1) first, the design must follow the national industry or other standards and must consider recent developments in standardized design solutions (see next chapters) and must provide the necessary technical accuracy that includes everything necessary, avoids unnecessary work and materials, and also considers a certain cushion, and (2) availability of accurate cost data: cost data is available on national level for different levels of design on a commercial basis. In Germany SirADOS is one of the most favored online tools in the commercial use. Also, BKI is a good source in this context. In a survey among 28 energy consultants [6] the majority stated that they would consider their own data collections as sources for investment cost estimation. In most cases, databases provide specific costs per floor space unit or per external wall and window area (windows, roofing, roof insulation, wall insulation, perimeter insulation), per system load (boilers, CHP units, biomass boilers); these costs comprise, in most cases, the material and labor cost in average.

Specific areas of concern for investment costs

The usage of investment costs must take into account several specifics:

- Seasonal variations of material and labor costs: experience shows that the application of insulation materials and adhesive airtightness materials can only be considered under certain climate conditions; the demand for the services (labor costs) is the highest in summer; cost savings often can be considered if the commissioning of material is considered in colder months.
- Regional variations of material and labor costs: the assessment of labor cost levels must be considered on the national and regional level.
- Energy- and non-energy-related investment cost: Investment costs should be distinguished with regard to their energetic relevance: Scaffolds, paint- and brush-costs, rendering and other non-energy-related investment costs and repurposing costs such as refurbishment of room equipment, sanitary installations etc.

2.4   Investment Cost: Reduction Potential in DER Projects

In the public sector, acceptance of DER projects is often related to the cost-benefit calculation, i.e., whether the project is cost- effective. As a first "rule of thumb," cost effectiveness is usually expressed as the coefficient of:

$$\text{Static amortization} = \frac{C_{investment}}{C_{savings}} \qquad (2.2)$$

The investment cost of DER projects still has a large optimization potential that must be exploited to increase the cost effectiveness of DER:

- **Economies of scale**: Best practice in a number of products (wind mills, PV panels, light emitting diode [LED] lighting systems) shows that increasing numbers of products will lead to significant cost reductions. On the level of a single DER project economy, the planner must consider the standardization of certain technical solutions and refer as much as possible to already existing technical solutions.

Design optimization: The economies of scale must work hand in hand with the optimization and selection process in the design of the products. In the LED sector, the reduction of different light temperatures and light intensities and the great variety of formats allow the DER to optimize specific cost intensive areas. Material and labor costs must be considered here. Research IEA EBC Annex 50 [7] show potential cost savings of 20% may be achieved by the use of prefabricated wall-roof window constructions. Another approach is the Energiesprong [8], which originated in the Netherlands, and which targets the reduction of investment costs by at least 30% through the use of prefabricated and well predesigned technology solutions for highly replicable building types. However, such cost-optimized designs must overcome specific technological weak points that tend to drive costs up, e.g., at the joints of different building components, where standardize solutions are needed to reduce design and labor costs on the construction site.

2.5   Optimization of Investment Scenario by Least-Cost Planning [9]

The cost structures of building energy retrofit, energy supply systems and energy distribution are interlinked. Therefore, an optimization is required that uses an iterative approach to account for the impacts of demand variations. This requires iterative steps to find the minimum of total costs. Since the annual energy demand will vary depending on the degree of ECMs, the decisive indicator to assess the cost effectiveness of different combinations of measures is not given by (minimized) energy costs (€/MWh), but by the resulting minimum of total (specific) energy costs (€/m^2) of the buildings under consideration.

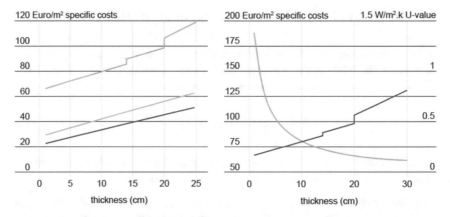

Fig. 2.1 Insulation costs, insulation thickness, U-values

In some modeling tools this cost optimization process is integrated. In the monthly modeling tools, however, the least-cost planning is a separate modeling activity to be provided by separate tools [10].

Funding should be used in a way that promises the largest impact on the decision-making criteria. In this case, this is the cost-benefit optimum of ECMs for buildings to reduce the heat losses through the envelope in several consecutive steps.

(a)   *Collect investment cost data:*

Figure 2.1 shows a typical structure of insulation costs for existing buildings [11].

On the left side: insulation costs per m^2 of insulated area, from below: basement ceiling (dark blue), attic bottom (light blue), external wall (orange) as function of insulation thickness (cm). On the right side: U-value of external wall (red curve and left vertical scale) as function of thickness (heat transfer coefficient $= 0.035$ W/mK) and insulation costs (blue curve and right vertical scale).

(b)   *Set up of modeling process:*

The heat transfer loss is directly proportional to the U-value. Figure 2.2, right, shows that the benefit of the additional insulation, the decreasing U-value, decreases with thickness while the costs increase more or less linearly. Figure 2.2 shows the heat losses over increasing thicknesses for a building in ASHRAE Climate Zone (c.z.) 5, 3,000 heating degree days and 20°C (68° F).

The upper curve is displayed without inclusion of embedded energy of the insulation material, the lower curve shows the results after subtraction of the embedded energy $e_e = 1,100$ kWh/m^3.

Figure 2.3 shows that beyond an insulation thickness of about 15 – 20 cm any additional insulation has hardly any effect. The upper limit of effectiveness for insulation thickness is about 23 cm in c.z. 5. This limit depends on the insulation material used, since different materials exhibit different embedded energy densities, varying from about 200 kWh/m^3 for insulation made of organic waste materials to 1,100

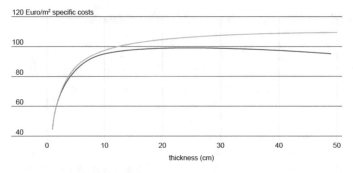

Fig. 2.2 Reduction of external wall heat transfer, $\Delta q_{thermic}/kWh/m^2 yr$, with increasing insulation thickness

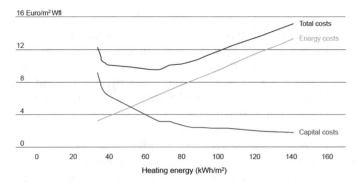

Fig. 2.3 Least-cost curve of ECMs for an existing multi-family building (30 flats, five floors, 2,000 m² living area)

kWh/m^3 for EPS. It further depends on the climate prevailing in the location of the case under consideration, since envelope insulation results in a higher energy conservation effect in colder climates (Fig. 2.2 was related to a local climate with 2,050 heating degree days).

(c) *Set up of least-cost planning pathway*

The least-cost path is achieved by a stepwise comparison of the capital cost and energy cost reduction of every ECM possible in the concrete case. For instance, increasing the attic insulation by 1 cm, results in an increase of capital costs, $\Delta c_{capital}$, and a decrease of the heating costs, $\Delta c_{heating}$. This is compared by the model with other possible ECMs, such as wall insulation. The measure yielding the largest ratio $\Delta c_{heating}/\Delta c_{capital}$ is chosen by the model in every iteration step. This numerical approach to find the least-cost curve provides results like those shown in Fig. 2.3.

Figure 2.3 shows the capital cost curve (bottom), energy costs (red straight line) and total heating costs (top); the chart must be read from right to left, beginning with $q_h = 140 \, kWh/m^2$ before retrofit.

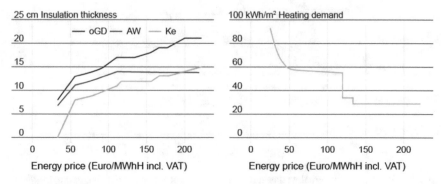

Fig. 2.4 Energy prices over insulation thickness and heating demand

The quantitative result of this model is a list of measures that contribute to the combination of measures resulting in minimized total heating costs (capital costs plus energy costs) of the considered building or building type. The simulation that led to Fig. 2.4 says that the cost minimum, assuming an end energy price of 8.5 ct/kWh (incl. VAT), is achieved in this example by an U_m-value (weighted average of envelope U-values) of 0.76 W/(m^2 K). In this case, window replacement is not part of the least-cost combination of measures.

The resulting heating energy demand is 66 kWh per m^2 living area (down from 140 kWh/m^2 pre- refurbishment), as indicated by the total cost minimum depicted in Fig. 2.4. The total costs of the refurbishment scenario include the capital costs of the selected ECM bundle and the remaining heating energy costs (upper curve in Fig. 2.4) which is 25% lower than the total costs, which are only the heating costs of the building before the refurbishment investment

The least-cost-path however is depending on the energy price level. Figure 2.4 is visualizing the results of the above mentioned simulation with different energy prices.

The left part of Fig. 2.4 shows the economically optimized values of the insulation thicknesses (left) and resulting dependency of the cost optimal heating demand, q_h, from the energy price level. The lines in the left chart belong to basement ceiling insulation (orange curve), external wall insulation (blue curve), and attic floor insulation (black curve), calculated for a climate characterized by 2,050 heating degree days and an internal temperature of 20°C (68°F). Also considered in the least cost calculation are three different insulation levels of windows (double pane, triple pane and triple pane at passive house quality). For external wall and basement, a saturation value for insulation thickness is achieved at about 14 cm at energy prices beyond 100 €/MWh. The right part of Fig. 2.4 shows resulting specific heating demand (kWh/m^2) corresponding to the total cost minimum of envelope and window retrofit measures following the least-cost pathway exhibited by the model as a function of varying energy prices (from 30 to 220 €/MWhHu, VAT (19%) included). The blue curve shows the impact of the different simulated measures on the specific energy costs; starting from left with most cost- effective measures at the basement ceil-

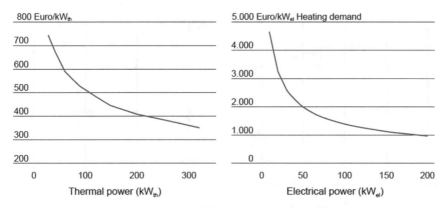

Fig. 2.5 Specific investment cost indicators for energy supply measures

ing, attic floor and external wall insulation until reaching the specific energy costs of approximately 125 €/MWh. The following step is the implementation of triple-pane windows which reduces the energy demand from 58 to 28 kWh/m²yr. After that step the specific energy costs may be increased by further measures which have no significant impact on the energy demand. In colder climates with more than 3.000 heating degree days this result would be different. But even in the case and location under investigation here, window replacement could be necessary because, after some 25 years of technical lifetime, windows airtightness is generally reduced and user comfort is impaired due to air draught and irradiation imbalance of the cold windows surface.

Figure 2.5 shows the size dependence of specific costs $k_{Pellets}$ of Pellet boilers (left) [12] and k_{cog}, the costs of cogeneration plants [13] ($k_{Pellets}$ in € per kWth and k_{cog} in € per kWel, respectively); in both cases the total investment costs are considered including installation, excl. VAT.

Optimization of Heating Supply

Energy performance is defined both by (optimized) energy demand and (improved) energy supply efficiency. Both effects, and the connected investment costs, eventually determine the total costs of the supplied energy. There is a direct connection between energy conservation and EEMs, since the investment costs of energy supply units depend on their size, as illustrated by the two examples in Fig. 2.6.

In the simple case of a monovalent pellets boiler with efficiency $\eta_{Pellets}$ and annual full load hours of h_f, and with pellets price $p_{Pellets}$ (€/kWh), the energy costs of this plant is given by

$$kth_{Pellets} = \frac{k_{Pellets} \cdot (a + w)}{h_f} + \frac{p_{Pellets}}{\eta_{Pellets}}(EUR/kWhth), \qquad (2.3)$$

where "a" denotes the annuity of the investment and "w" the maintenance costs as a fraction of the investment costs. Figure 17 illustrates this cost function of heat

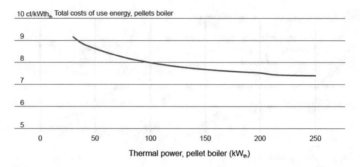

Fig. 2.6 Pellet boiler costs of heat generation

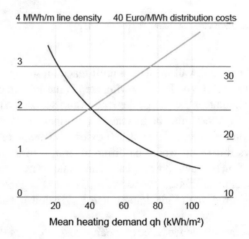

Fig. 2.7 Impact of distribution network density on the total distribution costs

generation inserting the pellets boiler costs displayed in Fig. 16, with $p_{Pellets} = 4.5$ ct/kWh, $n_{Pellets} = 0.82$, $w = 2\%$ of the investment costs per year, and $h_f = 1,700$ h/a. The annuity "a" is calculated using a depreciation time of 25 years with an interest rate of 4%.

District heating distribution networks: In the case of neighborhood scale heat distribution networks, the costs of the heat supply grid are a significant component of the total energy costs. While the specific costs decrease with the size of the heating central, the additional distribution costs must be included and a cost optimum must be found. Figure 2.7 shows the increase of the specific heat distribution costs (€/MWhth) in a case study [14] assuming a continuous heating demand decrease due to ECMs for an increasing number of buildings in the supply area, by which the "line density" (MWh/m) of the distribution network is reduced.

Figure 2.7 the orange curve and the left scale show the impact of decreasing distribution network line density (MWh/m distribution network line resulting from decreased mean heating demand q_h. The orange line relates high mean heating demand on the right side with high distribution line density in a linear function; by reducing the mean heating demand q_h by 50%, the line density is reduced by 50%. The impact of the level of mean heating demand on the distribution costs is displayed in the black curve and the right scale: high mean heating demand on the right side of this curve means low distribution costs of 15 €/m^2 distribution line, while half mean heating demand increase the distribution costs drastically.

The least cost planning process provided a cost- benefit optimized bundle of energy efficiency and energy supply measures which improved cost-effectiveness of the DER in comparison to the national minimum requirements by more than 20%. The conclusion of the case study was that the economic optimum of the combined costs of energy conservation, energy supply, and energy distribution network was a heating standard after retrofit in the range of 45 kWhth/m^2 compared to the average pre- renovation heating demand of 140 kWhth/(m^{2}yr) before retrofit).

References

1. Procedures for Commercial Building Energy Audits, ASHRAE 2011, Atlanta
2. IEA Annex 61. 2017 Deep Energy Retrofit – A Guide to Achieving Significant Energy Use Reduction with Major Renovation Projects
3. IEA Annex 61. 2017. Deep Energy Retrofit – Case studies
4. IEA Annex 61. 2017. Deep Energy Retrofit Pilot Projects
5. Energy Efficiency the First Fuel for the EU Economy, EEFIG, Brussels 2015. http://www.unepfi.org/fileadmin/documents/EnergyEfficiency-Buildings_Industry_SMEs.pdf4
6. Appelt, Lohse. INEECO Interim Project 2016/1, 2016, Karlsruhe. www.ineeco.org
7. IEA Annex 50, Final report, Zürich, (2012)
8. www.Energiesprong.nl
9. Reinhard Jank et al., *Deep Energy Renovation Optimization Model* (Ravensburg, Berlin, 2017)
10. Jank, Korb. DEROM Handbuch (Berlin/Ravensburg, 2017)
11. IEA Annex 51 Energy Efficient communities (2011–2014). www.annex51.org
12. internal report: Volkswohnung Karlsruhe (2010)
13. Municipal Administration of Frankfurt. Characteristic data on cogeneration plants, market study, Frankfurt (2011).
14. EnEff Stadt Research Project Rintheim. https://projektinfos.energiewendebauen.de/landing

Chapter 3
Evaluation of Cost-Effectiveness of DER Projects

From the perspective of ESCOs and financiers, not much information is available on the decision-making criteria in the public real estate sector. This chapter describes the basic definitions and structures of cost calculation and cost effectiveness calculation in the public real estate sector. These calculations are necessary for the better understanding of DER business models and financial models needed.

3.1 Cost Effectiveness Calculation in the Real Estate Sector

In the public sector, buildings are managed using more or less similar KPI and decision-making structures that ensure the cost-effective delivery of the building and its functionalities over the life cycle of the building and its components. The following sections describe cost effectiveness criteria currently in use, in the context of DER projects in the public building sector.

Cost effectiveness of a public real estate object

From the perspective of a DER project, a real estate object [1] is seen as cost-effective when it generates "positive" income (Table 3.1). Similarly, the payback of investment plays a major role for public building agencies that provide their users with DER refurbished buildings. The "split-incentive" problem occurs when the public agency cannot allocate the full investment costs into the generated revenues. One reason for this is that the usage fees or rental rates provided by the public users often consider only the average rental rate of the fair market values [2].

© The Author(s), under exclusive license to Springer Nature Switzerland AG 2019
R. Lohse and A. Zhivov, *Deep Energy Retrofit Guide for Public Buildings*,
SpringerBriefs in Applied Sciences and Technology,
https://doi.org/10.1007/978-3-030-14922-2_3

Table 3.1 Cash flow and EBITS in real estate sector

		Rental income, energy production
–	Gross revenues	Rental income, energy production
–	Costs related to the gross venues	
=	Gross revenue total	
–	Operational costs:	Maintenance and repair, administrative costs, insurance, operational costs not to be recovered from tenants
=	Net revenues (CASHFLOW)	
–	Refurbishment costs	DER costs, general refurbishment (payback)
=	Operating result (EBIT)	
–	Taxes and liabilities	Taxes on EBIT and asset values, financing costs (interest rates)
=	Income	

3.2   Life-Cycle Cost Calculation

An evaluation of public decision-making in the building sector shows that the most decisions are made on the basis of investment costs [3]. However, operation costs over the life cycle of a building are many times greater than the initial construction costs [4]. Decisions made in the early stages of the programming and design process influence LCCs in terms of space, the quantity of structural elements, technical and mechanical service equipment, and the choice of materials.

The standard method to analyze the project cost effectiveness is to perform a life cycle cost analysis (LCCA), which accounts for present and future costs of the project. LCCs in buildings are defined by the standards on Life Cycle Cost Analysis ISO 14040-44, ISO Standard 15686-5, the European standards EN 15804 and EN 15978, and other reference documents of the ILCD 2010a, ILCD 2010b, ILCD 2010c and more specific and operative approaches such as German Industrial Standard VDI 2067, B1 [5]. In the United States, energy efficiency of Federal construction and renovation projects shall be evaluated using Federal Life Cycle Cost Methodology and Procedures outlined in Handbook 135 using BLCC 5.3-09 software package (10 Code of Federal Regulations [CFR], Part 436). Life Cycle Costing is a tool and technique that enables comparative cost assessments to be made over a specified period of time. The tool takes into account all relevant economic factors both in terms of initial capital costs and future operational and asset replacement costs, through to end of life, or end of interest in the asset. The tool also takes into account any other non-construction costs and income, which are defined in the scope [6].

Scope of LCCA: The way that the economic and ecological goals are set will influence the efficiency and effectiveness of an appropriate design that considers the LCCA. The decision-making process that steers a building refurbishment must make a full life cycle assessment to direct the process to achieve energy, embedded energy, waste, and recycling and disposal goals. However, the DER decision-making process focuses on the assessment of Part-LCC, which considers the construction and usage phase of the building; it does not focus on the deconstruction phase.

Definition of LCCA investment cost breakdown:

In LCCA, one must select the base case scenario that will be used to evaluate the cost effectiveness of the DER. Most of major renovation projects include a scope of work, which can be either energy- or non-energy-related. A non-energy-related scope of work may include elements such as different construction jobs related to changing floor layouts (e.g., moving/removing internal partitions), adding bathrooms, removing asbestos, adding a sprinkler system.

An energy-related scope of work of major renovation project typically includes replacement of existing mechanical, lighting and electrical systems, replacement of some or all windows, and replacement of existing ductwork and plumbing systems. A major renovation with an energy-related scope of work that will meet current minimum standard requirements will be considered as a base case for the LCCA.

While a non-energy-related scope of work will remain the same in both the base case and DER scenarios, the energy-related scope of work will be using higher efficiency equipment and systems, and will consider additional measures, e.g., building envelope insulation, improvement of building air tightness. Some improvements, such as additional insulation, high performance windows, will increase the cost of renovation; other improvements (e.g., smaller heating and cooling systems, boilers, and chillers) can reduce costs of renovation compared to similar budgeted items. Therefore, the overall budget for a DER project is typically higher than the cost of a major renovation that follows only minimum energy requirements. However, the reduction of heating, cooling, or humidity loads resulting from implementation of DER measures will result in a smaller, sometimes simpler HVAC system, which in turn reduces both initial investment and capital replacement costs related to these systems (Fig. 3.1).

Period of LCCA: The period of LCCA is the length of time over which an LCCA is analyzed. This period of analysis shall be determined by the building owner at the outset of the project, on the basis of the probable life cycle of the asset itself, or on the basis of specific national regulations [7].

Scope of costs: The LCCs typically include the following two cost categories: investment-related costs and operational costs.

Investment-related costs include costs related to planning, design, purchase, and construction. The selection of the investment costs sources may have a large impact on the reliability of the LCCA: Three main sources for data for LCCA purposes are: (1) from the manufacturers, suppliers, contractors, (2) historical data, and (3) data from building modeling databases. Data from existing buildings are used as historical data; some of them are published in the BMI (Building Maintenance Information) occupancy cost databases available in a couple of countries for the national data. Other data sources include clients' and surveyors' records, and journal papers. Existing

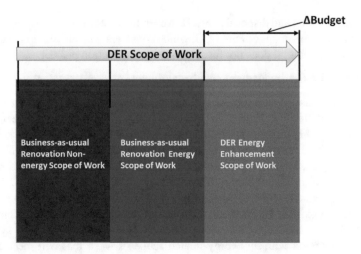

Fig. 3.1 Scope of work of DER project

databases have their limitations in that they do not record all necessary context information about the data being fed into them. The data are usually expressed as units of cost, which limits them to local use [8].

Initial or first investment costs describe the total expenses of the initial DER investment. It includes planning, modeling, design, implementation of new materials, and replacement and disposal costs of replaced materials, including both material and labor costs. The number and timing of **capital replacements or second investments** depend on the estimated life of the system and the length of the service period. Sources for cost estimates for initial investments can be used to obtain estimates of replacement costs and expected lives. A good starting point for estimating future replacement costs is to use first investment cost along with increasing factors related to the national inflation rate of comparable building construction and HVAC investment cost indexes [9].

Synergetic impacts: The determination of the first investment costs must consider the timeline of the DER implementation. Assuming that the implementation will be carried out in a one-staged process, first investment cost reductions will be achieved by downsizing the HVAC, especially the supply, and by ventilative heating and cooling resulting from a thermal envelope improvement, or the implementation of a building management system (BMS). The modeling process must consider and evaluate these measures.

Also, combining a DER with a major renovation will result in smaller investment costs compared with performing a DER on its own. When a DER is combined with a major renovation, "anyway" costs for building evacuation and gutting, installation of scaffolding, replacement of single-pane and damaged windows, or even replacing and/or upgrading building envelope insulation to minimum standards are already a part of the major renovation project and need not be accounted as an additional cost for related to the DER.

Grants: Grants, rebates and other financial subsidies for energy efficient and sustainable design (one-time payment) will reduce the first investment costs. In European countries, major grant programs provide grants for partial or holistic renovation scenarios with regard to the incremental investment costs compared to the national minimum requirements. Rates vary from country to country in a range of 20–50% of the incremental DER investments.

The **residual value** of a system (or component) is its remaining value at the end of the study period, or at the time that it is replaced during the study period. Residual values can be based on value in place, resale value, salvage value, or scrap value, or on the net of any selling, conversion, or disposal costs. However, as a general rule of thumb, the residual value of a system with remaining useful life in place can be calculated by linearly prorating its initial cost. For example, for a system with an expected useful life of 15 years that was installed 5 years before the end of the study period, the residual value would be approximately 2/3 [= (15–5)/15] of its initial cost. Comparable to the ISO 15686-5 USDOE FEMP LCC methodology requires that **residual values** (resale, salvage, or disposal costs) and capital replacement costs are included as investment-related costs. Capital replacement costs are usually incurred when replacing major systems or components, paid from capital funds.

Operational costs: An economic evaluation of a DER usually considers only energy costs. The current state of the art in LCCA of public and commercial buildings also considers the following operational costs:

1. Maintenance, operation and management are necessary for ensuring that a building functions and operates properly throughout its life cycle including regulatory maintenance costs, e.g., repairs, replacement, refurbishment; The maintenance activities usually include inspection, monitoring, testing, condition inspections, maintenance planning, repairing, refurbishing, and partial replacements. The following indirect impacts of maintenance works can also be taken into account: down time (loss of function for a period).
2. Insurance costs for building hazard, fire protection, pipe work, electric installation.
3. Energy, water and sewage costs.
4. Revenue from ownership or use of the asset.
5. If necessary further operational costs may be considered, e.g., facilities management, annual regulatory costs (e.g., fire, access inspections) and demolition, cost of disposal, unanticipated costs resulting from legislation introduced subsequent to completion of the constructed asset, e.g., in relation to environmental, health and safety requirements or fiscal matters.

Calculated for each scenario, in addition to the energy benefits also the non-energy benefits the following cost-reductions compared to the baseline scenario should be considered:

1. Energy use and cost reduction due to improved efficiency of the building and its systems
2. Energy cost reduction due to shifting energy peaks, switching to different fuels (e.g., using cogeneration or tri-generation) or replacing fossil fuel based thermal or electrical systems to systems from renewable energy sources
3. Maintenance cost reduction with replacement of worn-out equipment at the end of its life cycle
4. Maintenance cost reduction due to downsizing of mechanical systems with reduced heating and cooling loads
5. Operation cost reduction using advanced building automation systems (BASs).
6. In some scenarios, energy use may increase compared to the Base Case due new requirements to indoor air quality or thermal comfort. For example, adding cooling or humidity control requirement will result in additional energy use for cooling systems. Maintenance cost of some replacement systems may increase due to the complicity of controls system, but may be offset by reduced energy use resulting in more efficient operation of the HVAC system:

   a. **Interest (discount) rates**: chosen interest rate related to the timeframe and the current market situation.
   b. **Price increase**: assumptions for the price increase; constant rate of growth may lead to unrealistically high energy prices for long calculation periods.
   c. **Capital costs**: usually take the largest part of the annual costs and are usually expressed as constant annuities over time. In projects with a timeframe greater than 20 years, the interest rates usually are not kept fixed; however, the write-off will, in most cases, be constant.

3.3   Cost Effectiveness of DER Investments

First costs are considered to be high and short-term investment horizons are considered to be unsatisfactory; in commercial and public sector the priority is to keep the buildings in the life-cycle and not to focus sustainable refurbishment and energy strategies beyond the minimum requirements, rental income, and short-term ROI. Thus, investment horizons that stretch beyond 15–20 years do not generally accommodate the long-term payback often needed to reach DER.

The cost effectiveness of real estate objects is assessed from several different perspectives:

- Financiers, banks, funds
- Building owners
- Building users, tenants

using variety of dynamic KPIs:

- NPV
- Net- and Gross ROI

- Cash flows
- Internal rate of return (IRR).

Static calculation methods (payback period, return calculation) are only used for first estimates.

3.3.1 Cash-Flow Analysis

The economic and financial LCCA is built on a dynamic cash flow model of the DER case study, with a focus on the perspectives of potential investors and financing institutions. For this purpose, the projected income and expense cash flows are modeled over the defined LCCA time period. Economic KPIs derived are the IRR, the NPV, and a dynamic amortization period, separately for the project (P-CF) and for the equity cash flow (E-CF). On the financing side, the influence of typical debt ratios of 70% on the remaining equity cash flow and liquidity is examined using the financial KPIs "Cash Flow Available for Debt Service" (CFADS) and the "Loan Life Coverage Ratio" (LLCR).

The analysis also includes a multi-parameter sensitivity analysis of the IRR and NPV with respect to deviations in relevant input parameters, e.g., investment expenditures (CAPEX), operating expenditures (OPEX), price development of the energy cost baseline and on other LCA and the project duration.

In the cash flow analysis, the annual costs and benefits are displayed under assumed price increase rates. The Net present value of the costs and benefits are compared to the total investment costs; the cost effectiveness is achieved if NPV − Investment costs > 0 within the Figs. 3.2, 3.3, and 3.4 [10] show a three-step cash flow analysis for a DER project over a time period of 30 years for a DER project (passive house building refurbishment) of an office building.

Figure 3.2 shows the operating costs (OPEX) for electricity, heating, and maintenance (each including an annual price increase of 2%) over a LCCA time period of 22 years.

Figure 3.3 shows the comparison of two scenarios Sc 7, −55% energy savings against baseline) and Sc 3, the implementation of a passive house. The accumulation of the savings (electricity and heating) over time corresponds to the "available" over all investment cost budget resulting from the operational cost savings (OPEX savings).

Figure 3.4 shows the OPEX is displayed over time, with a significant reduction in 2015 for both scenarios and the cumulated OPEX savings as the NPV. The NPV shows the equivalent investment that can be balanced by the OPEX savings.

The present value of revenue is the amount needed today to yield the same revenue from the bank, including interests. The present value of expenses is the amount currently needed to pay upcoming expenditures. Present values are comparable as they refer to the same point in time. For effective economic assessments, it is useful to do the calculation based on real prices and interest rates. Inflation (which does not affect the economic result) is not considered.

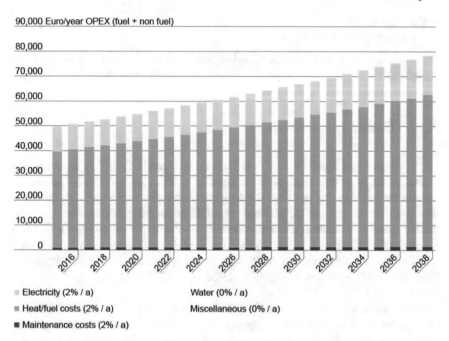

Fig. 3.2 Projected OPEX for the base case of a DER project

The NPV is the sum of all present values. Costs (or payments, e.g., the investment) are negative and revenues are positive. The NPV is the total gain of the investment when all lifetime costs and revenues are taken into account. Therefore, a positive or non-negative NPV means that the investment is economical. As long as capital (incl. debt) is available, it is economically profitable to make any investment up to a NPV of 0. NPV is a measurement of the profitability of a DER investment by subtracting the present values of cash outflows (including initial cost) from the present values of cash benefits over a period of time. A positive NPV means that the investment is balanced by the benefits.

$$NPV = -I + SUM(n;t)\frac{ct}{(1+i)^t} + \frac{RV}{(1+i)^n} \tag{3.1}$$

NPV Net Present Value function
I   DER investment in year $t = 0$
t   time period
n   study life in years
ct   cash flow in year t
i   interest or discount rate
RV   residual value in the year n.

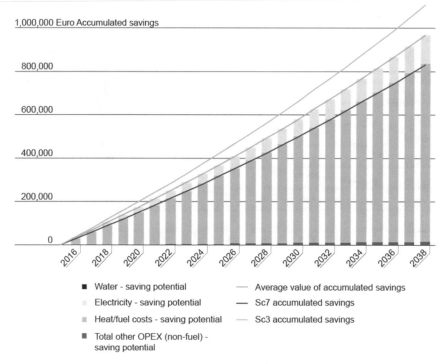

Fig. 3.3 Projected accumulated energy savings from DER project scenarios (Sc7: −55% energy savings, Sc3: Passive House)

Application of the NPV in DER projects: Comparison of DER to base case scenario

A DER enhancement to a major renovation project will be cost effective when the budget increase allowance, compared to the Base Case scenario designed to meet minimum energy requirements, will not exceed the NPV of all operating costs reduction and the increase in revenues. The potential investment costs, which can be funded on the basis of the life-cycle costs, can then be calculated using the formula for NPV calculation as:

$$\Delta \text{Budget}_{\text{max}} = \text{NPV}\left[\Delta \text{Energy(costs)}\right] + \text{NPV}[\Delta \text{Maintenance(costs)}]$$
$$+ \text{NPV}\left[\Delta \text{Replacement Cost (costs)}\right] + \text{NPV}[\Delta \text{Lease Revenues (costs)}]$$

$$(3.2)$$

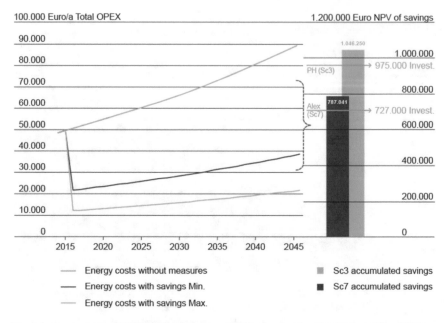

Fig. 3.4 Comparison of NPV and investment costs of both scenarios Net Present Value (NPV)

3.3.2   Internal Rate of Return-IRR

The IRR is a metric used in capital budgeting to measure the profitability of potential investments. IRR is a discount rate that makes the NPV of all cash flows from a particular project equal to zero. IRR calculations rely on the same formula as NPV. IRR are considered to compare several DER concepts:

$$-I + SUM(n;t)\frac{ct}{(1+i)^t} + \frac{RV}{(1+i)^n} = 0 \tag{3.3}$$

I   DER investment in year t = 0
T   time period
n   study life in years
ct   cash flow in year t
i   interest or discount rate
RV   residual value in the year n.

3.4 Multiple Benefits: Bankable LCC in DER

While a standard building LCCA considers a large scope of operational costs, the majority of DER cost-effectiveness calculations consider only energy cost savings. An evaluation of DER projects [11] has shown that many average static payback periods are often longer than 25 years if energy cost savings only are considered.

However, DER investments often produce benefits beyond reduced energy consumption and peak demand shaving. Many of these benefits contribute to the objectives of organizations implementing the projects and can have significant added value for those making investment decisions [12]. Although research has been carried out on specific topics such as the impact of increased thermal comfort on the productivity of the building occupants, or the willingness to pay increased sales prices or rental rates, the monetarization of the non-energy benefits, or "Multiple Benefits" (MBs) [13] is still not broadly used. The first step to provide a systematic assessment of MBs is to list and classify potential benefits using their potential impact, the primary beneficiaries, first approaches for the monetarization, and the way that the M&V process can be conducted. It will be easier to monetarize MBs pertaining to costs and benefits that have already been explored and made accountable in the context of building LCCA, and that provide M&V schemes.

Methods of quantification vary widely between benefits, and depend on the desired accuracy of financial estimates. As yet, there are no standards for quantification. The concept is still evolving. Benefits are being studied in different applications and methods are still being developed.

An important requirement for the MB is its relevance to project financing. In other words, a benefit should be considered part of the equity rate that is necessary to gain access to a bank loan or other third party financing. In a financial assessment of a project, this means that savings are considered to be a revenue source, which can then be considered on the equity side of project.

As the experience with EPC [14] shows, any kind of benefit may be considered in the financing scheme if: (1) it can be measured and verified (M&V) on a transparent way and (2) one participant is providing a performance guarantee. This is usually the case within an EPC project, in which the ESCo provides both conditions.

Figure 3.5 shows the relation between risk premiums and the transparency of the M&V process. For example, health effects are often discussed as an important side effect of a DER, however, these effects are hardly measurable. In general, the more subjective and the less transparent the M&V process, the higher the risk of disputes between the parties, the lower the bankability, and the higher the risk premium, which is usually expressed by the level of interest rate. According to the definitions of the globally used IMPVP [15] protocols an objective M&V process requires the comparison of established (i.e., agreed upon by the building owner and the parties carrying out the DER) baseline and post-DER part life-cycle costs (DER PLC) normalized to reflect the same set of conditions.

The pre-retrofit energy usage baseline (see step 2: baselining) is the starting point for an appropriate M&V process. The standard method is to utilize the original

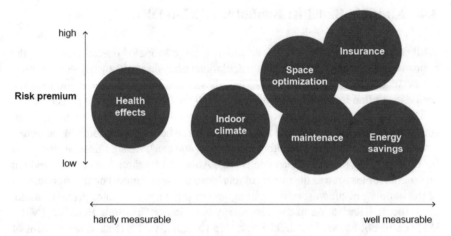

Fig. 3.5 Relation between risk premium and measurable impacts

regression-driven baseline model, and to apply it to post-installation conditions to represent what the baseline energy use would have been in the absence of the DER, and to its energy conservation program in the building (IPMVP Option C). Savings are determined by comparison to the established baseline energy and post-installation energy use, adjusted to the same set of conditions. The approach requires adjustments to baseline energy use to routine adjustments (expected changes such as weather normalization for the heating demand) and non-routine adjustments (occupancy, type of space use, operating hours, service levels such as increased indoor temperatures due to a new purpose of floor space for the energy consumption).

3.4.1   Bankability and Risk Mitigation of Multiple Benefits

To support bankability, the risks of MBs must be evaluated in a sensitivity analysis. The major risk of the economic calculation of a DER project can be aligned to the weight of each cost position in the LCCA. In DER projects, 50–80% of the LCCA costs are capital costs. Thus, in the greatest risks are the inaccuracy of the investment costs and the annual capital costs. For example, the investment costs could be negatively influenced by an early shortfall (before the end of the life time period under consideration) of a component or system. This would require secondary investments at an earlier point of time than calculated in the LCC cash flow, a higher write off, and additional capital costs for the secondary investment.

However, the assumption on the energy savings could be inadequate. To mitigate such risks, the cost estimation and the sensitivity analysis consider a minimum and a maximum mismatch value (Table 3.2). The risk mitigation will consider the level of cost impact risk on the LCA cash flow.

Table 3.2 Risk evaluation in a sensitivity analysis

Risk	Defaults	To be considered in the sensitivity analysis
Primary and secondary investment costs	Mismatch of investment costs, default in price and or m² calculation	Spread of investment costs for primary and secondary investment costs
Capital costs	Early default of primary investment, not adequate interest rate	Spread of interest rates and reduced life time for major primary investment
Energy-saving predictions	Energy performance is not equivalent to energy modeling	Spread of different savings
Operation costs	Not adequate estimated	Spread of different operation costs

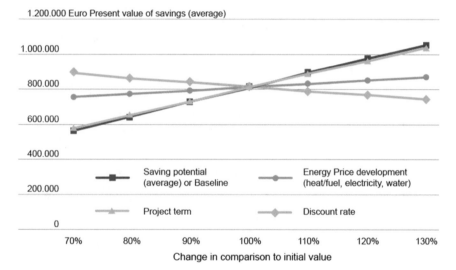

Fig. 3.6 Impact of different defaults on the NPV of savings

The sensitivity analysis of the case calculated in Fig. 3.6 assessed the impact of different defaults on the NPV of savings. The default of energy savings (the baseline) has a 1:1 impact on the NPV: 10% less energy savings may reduce the NPV by 10%.

An assessment of several case studies carried out in the United States and Europe [16] (along with discussions with public building owners) determined how multiple benefits may be considered as part of an advanced part life-cycle cost analysis to improve the cost effectiveness of a DER (Table 3.3).

Table 3.3 Multiple benefits and their values

	Part-LCC	Calculation method	Variations and values
1	Energy savings: effects from improving the e-performance	kWh savings × energy price	Fixed or flexible energy price; in DER it is expected to at least reduce by 50%
2	Energy savings II	kWh RE replacing fossil × energy price (RE-fossil)	kWh replaced by RE; fixed or flexible energy prices
3	Reduced maintenance I	Maintenance costs for replaced, worn-out equipment at the end of its life cycle as a percentage of the new investment value	Average percentage value or end of life cycle value
4	Reduced maintenance II	Downsizing of investment in a DER bundle means reduction of investment cost related maintenance	A component downsized by 30% reduces maintenance costs of this component
5	Reduced operation costs	Building automation reduce operation workloads	Consider work plans and operation schedules individually
6	Insurance costs I	Building components replaced achieve lower premiums and improved protection against loss	EU: compared to pre-refurbished status: –2 up –4€/m^2
7	Rental costs for floor space	m^2 savings x rental rate	DER may contribute to more flexible room concepts and certainly downsize space needs for mechanical systems
8	Reduced absence costs	Relationship between indoor climate, lighting, and absenteeism	Few case studies assessed the relationship: 30–40% less absenteeism

3.5 Energy Savings in DER

Improvements to a building's energy performance by the use of thermal insulation, heat recovery, and improved supply and demand appliances such as new boilers, ventilation, and high efficiency air-conditioning will reduce energy consumption by a certain amount:

$$\text{Energy Usage New} = \text{Energy Usage Baseline} + / - \text{Adjustments} \qquad (3.4)$$

Assessment of utility bills:

To establish energy consumption and cost baseline, utility bills are evaluated over a time span of 36 months (60 months in the case of oil usage). In this time period, trends must be assessed to develop a realistic baseline:

- a stable consumption can be baselined by the average of 3 (or 5 years)
- increasing consumption: the consumption of the last year is relevant
- decreasing consumption: the consumption of the last year is relevant.

Climate adjustment of the energy baseline

The heating part of the gas, oil or other heating site energy consumption must be corrected by heating degree days. For this, variable base heating degree day correction factors are available for each year. In most cases the correction will be carried out by a long term heating degree value of an average (national) region of the climate zone. The IMVP protocols consider the calculated ambient temperature for heating demand in the building, the average indoor temperature, and the long term average Kelvin days.

The process part of the heating and electricity consumption must be adjusted to those verifiable, variable parameters that influence these consumptions. For example, in a swimming pool, the impact of variable number of guests has an impact on the fresh water, the pool water heating consumption, and the electricity consumption of the pool water circulation pumps.

Energy baseline prices:

Energy savings will be calculated against the baseline using recent energy prices (described in Chap. 5):

- **Energy consumption baselining**: The first step is to define the energy baseline (see above) as a verified basis to calculate, monitor and verify the energy savings. Baselining includes heating and cooling degree adjustment; consumption related to 365 days/year and a "normalized" operation and usage of the building in a way that eliminates abnormal disturbances such as construction, hazards, etc.
- **Energy price baselining**: To monetize the value of energy savings, current or future price scenarios for existing supply solutions and RE are taken into account. Energy tariffs consist of fixed partitions (annual fixed price, meter price, load price) that are not volatile as long as the building gets disconnected from this energy source (power, district heating, district cooling or energy supply contracting). The cost value to be considered in the case of "normal" energy savings is the price per kWh or BTU/h at the utility meter and fees related to the energy consumption.
- **Energy Price development**: Long term DER payback periods must consider a price increase rate; fixed annual increase rates include the risk that, over time, rates might actually differ so that the monetary value might become either too high (which will be a problem for the building owner) or too low (which will unnecessarily increase the payback period).

Table 3.4 Calculation of energy cost savings from energy savings and fuel switching

	Heating	Energy savings	Fuel switch
Energy/price baseline	100,000 kWh/year	0.06 €/kWh	0.03 €/kWh
Energy saving I	50,000 kWh/year	3,000 €/year	
Energy saving II (fuel switch)	40,000 kWh in RE		1,200 €/year
Total	4,200 €/year		

- **Floating or fixed energy price assumption**: Annual changes in energy prices, either as statistic indicators or building-related energy purchase prices are of critical interest to the investor. However, these "floating prices" creates risks to the investor that the investor (ESCo, building owner) cannot control; price decreases can severely reduce the cost-effectiveness of the project over time. Also, energy price adjustments may occur once or twice each year, which creates an additional workload for the performance M&V process of the project. In most U.S. and European ESCo projects, the energy price is kept either constant, or at a constantly increasing level over time. However, this assumption must be considered in the context of national energy market prices.
- **Case: Energy price risks with fixed energy prices in recent years**: Risk of failure did not exist from the perspective of in mid-term periods in Germany:

  - The German index for natural gas (>10,000 m³/year) moved from 54 in the year 2000 to 125 in the year 2015 with two curbs in 2006/7 and 2009 (Fig. 3.7a); for electric power, the index moved from 73 to 116.
  - In a project with 1,000 MWh natural gas savings between 2003 and 2015, actual value on actual prices averaged 613T€ compared to a fixed price value (2003) of 408 T€; a value of 408 T€ was collected from energy savings; the price increase between 2003 and 2015 added value of 205T€ (~50%).
  - In the same case, with 1,000 MWh el power savings between 2003 and 2015, with a value at a fixed price of 934T€, the added value after a price increase was 321T€ (~28%).

- Multiplying savings by the price allows an appraisal of the energy savings and the value of fuel-switching against the total energy cost savings (see Table 3.4), defined by:

$$\text{ECS} = \sum_{\text{year 1}}^{n} (E_{\text{savings}} \times \text{FEP} \times PIR) + \left(E_{consump., replaced} \times (EP - RE\,EP)\right) + FEC$$
$$+ (FIT \times FIA) \tag{3.5}$$

(a)

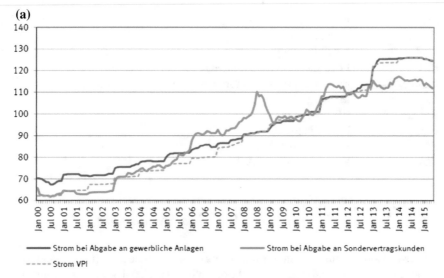

Legend: blue line — price index for commercial plants, blue dotted line: small
consumer power price index, green line- power price index for industry
consumers > 5 GWh/a

(b)

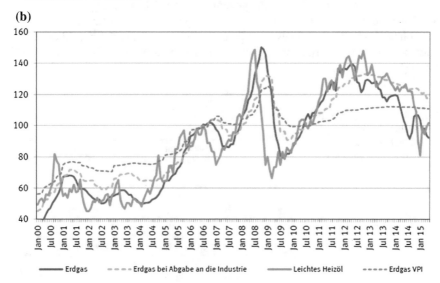

Legend: blue line- natural gas for small consumers < 1GWh/yr; blue dotted line-
natural gas industry price index > 5 GWh; green line- oil price index < 1 GWh/yr;
green dotted line: natural gas price index for small consumers < 1GWh

Fig. 3.7 a Power price index (above) Fig. 3.14b. Natural gas (below) price index for industry in
Germany 2000–2015

ECS	Energy cost savings
$E_{savings}$	Energy savings
FEP	Flexible energy price
PIR	Price Increase Rate (1/year)
$E_{consump., replaced}$	Energy consumption replaced by RE or CHP source
EP	Energy price
RE EP	RE Energy price
FEC	Fixed Energy Costs
FIT	Feed in Tariff (KWh)
FIA	Feed in Amount (kWh).

3.5.1   Additional Benefit for CHP: Grid Stabilization

Revenues from the interaction of timed energy demand reduction (i.e., peak cutting by demand reduction and storage) and the operation of regional power (i.e., frequency stabilization in case of power dips) is mostly unexploited in most life cycle assessment calculations reduction. As small generator-sets with synchronous generators typically have a comparably small inertia, the generator will accelerate fairly fast because of a voltage dip in the power grid within in the range of 150 ms to 1.5 s depending on the voltage level. The revenues must be collected from the regional grid operating company, but can in certain areas provide attractive business models for CHP [17] fleets (Fig. 3.8).

Fig. 3.8 LVRT requirements for synchronous generators, according to Germany's medium-voltage grid code [18]

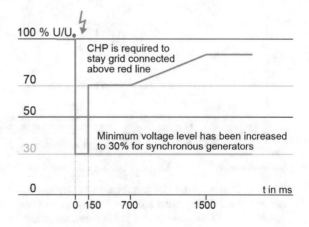

3.6 Avoided Maintenance and Repair Costs

Effective maintenance and repair is one of the most cost-effective methods for ensuring reliability, safety, and energy efficiency. Inadequate maintenance of energy-using systems is a major cause of energy waste in both the public and private sectors. Energy losses from steam, water and air leaks, uninsulated lines, maladjusted or inoperable controls, and other losses from poor maintenance are often considerable. Good maintenance practices can generate substantial energy savings and should be considered a resource. On the other hand, the public sector tends to cut costs, which results in poor maintenance (which results in degradation of building systems performance), and a drift of environmental sensors such as temperature sensors away from their setpoints, which may cause poor operations and increase in energy consumption [19].

The operation and maintenance of public buildings must comply with national standards and requirements. Table 3.5 lists some examples.

A bundle of technical measures carried out within a DER can support and partly subsidize the public building owner´s obligations to maintain and refurbish the building and to ensure the functionality and agreed levels of comfort and energy efficiency. The maintenance and repair in public buildings is based on a number of standards and requirements.

Table 3.5 Overview on current regulations in selected countries

Country	Definitions	Specific regulations and working material for building refurbishment
Germany	DIN EN 13306, DIN 31051 (2006)	Heating systems: BDW Information sheet 14; DHW: DIN 1988 T8, DIN 18960 Costs in civil construction; INQA Bauen "Instandhaltung von Bauten und technischer Gebäudeausrüstung" [20]
Switzerland	SIA 469 (1997)	CEPE ETH Zürich, IP Bau Kostenplanung I (1994), IP Bau Grobdiagnose, MER Habitat (quick evaluation of building conditions)
United States	General Services Administration (GSA) PBS P100 (2014), CFR Title 40, Protection of Environment, Executive Order 13423—Strengthening Federal Environmental, Energy, and Transportation Management, and Executive Order 13514, Federal Leadership in Environmental, Energy, and Economic Performance	Energy: Leadership in Energy and Environmental Design (LEED) Gold standard (sustainability, energy and environmental design), Green Building Ratings (U.S. Green Building Council), EPAct 2005, EISA 2007, ASHRAE 90.1R; Life Cycle costing: CFR Title 10
ISO	13823	Definitions

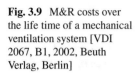

Fig. 3.9 M&R costs over the life time of a mechanical ventilation system [VDI 2067, B1, 2002, Beuth Verlag, Berlin]

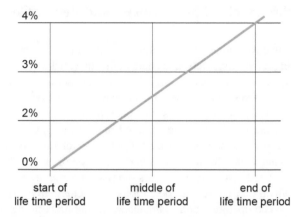

Typically, M&R measures are low- or no-cost in nature. It has been estimated that M&R programs targeting energy efficiency can save 5–20% on energy bills without a significant capital investment, especially when energy retrocommissioning is a part of the M&R program [21]. Maintenance and replacement costs change over the life time of building systems (see Appendix).

Based on empirical data of M&R costs of HVAC systems Fig. 3.9 shows in a simplified way how these costs vary over time as a percentage of the installed new equipment costs.

Overview on M&R costs.

M&R are defined as decisions and actions regarding the control and upkeep of property and equipment. These are inclusive, but not limited to:

- actions focused on scheduling, procedures, and work/systems control and optimization; and
- performance of routine, preventive, predictive, scheduled and unscheduled actions aimed at preventing equipment failure or decline with the goal of increasing efficiency, reliability, and safety.

Repairing an asset lowers the cost of the maintenance work and extends the life of the system or component. Identifying facilities management practices that plan, schedule and perform on-time routine and preventive maintenance provides the greatest ROI [22].

The USDOE FEMP Program's Operations and Maintenance (O&M) Guide [23] offers a number of metrics that can be used to evaluate an O&M program. Not all of these metrics can be used in all situations; however, a program should use of as many metrics as possible to better define deficiencies and, most importantly, to identify relevant potential improvements within a DER.

- **Capacity factor**—Relates actual plant or equipment operation to the full-capacity operation of the plant or equipment. This is a measure of actual operation compared to full-utilization operation.

Table 3.6 Industry R and M metrics and benchmarks

Metric	Variables and equation	Benchmark (%)
Equipment availability	$\% = \dfrac{\text{Hours each unit is available to run at capacity}}{\text{Total hours during the reporting time period}}$	>95
Schedule compliance	$\% = \dfrac{\text{Total hours worked on scheduled jobs}}{\text{Total hours scheduled}}$	>90
Emergency maintenance percentage	$\% = \dfrac{\text{Total hours worked on emergency jobs}}{\text{Total hours worked}}$	<10
Maintenance overtime percentage	$\% = \dfrac{\text{Total maintenance overtime during period}}{\text{Total regular maintenance hour during period}}$	<5%
Preventive maintenance completion percentage	$\% = \dfrac{\text{Preventive maintenance actions completed}}{\text{Preventive maintenance actions scheduled}}$	>90
Preventive maintenance budget/cost	$\% = \dfrac{\text{Preventive maintenance cost}}{\text{Total maintenance cost}}$	15–18
Predictive maintenance budget/cost	$\% = \dfrac{\text{Preventive maintenance cost}}{\text{Total maintenance cost}}$	10–12

- **Work orders generated/closed out**—Tracking of work orders generated and completed (closed out) over time allows the manager to better understand workloads and better schedule staff.
- **Backlog of corrective maintenance**—An indicator of workload issues and effectiveness of preventive/predictive maintenance programs.
- **Safety record**—Commonly tracked either by number of lost-time incidents or total number of reportable incidents. This is useful in getting an overall safety picture. A DER will be able to target especially unreliable equipment and to use this information to drastically improve the safety record, reduce lost hours, etc.
- **Energy use**—Energy use is a key indicator of equipment performance, level of efficiency achieved, and possible degradation. Energy use is commonly expressed in benchmarks of end energy use per floor space unit and site energy per fuel unit.
- **Inventory control**—Older technical equipment (e.g., electric drives, building automation) is often not supplied with spare parts after a life-time period of 5 years. To maintain the availability of these systems requires the purchase and storage of spare parts. An accurate accounting of spare parts can be an important element in controlling costs. A relevant consideration for DER projects is an accounting of the reduction in the need for (and cost of) outdated and unavailable M&R spare parts that are no longer needed when the relevant machinery is replaced in a DER.

The metrics and benchmarks listed in Table 3.6 can serve as a guide for evaluating the current repair and maintenance activities [24].

Cost-Optimization strategies:
The need for maintenance is predicated on actual or impending failure—ideally, maintenance is performed to keep equipment and systems running efficiently for at least the design life of the component(s). As such, the practical operation of a component is a time-based function. If one were to graph the failure rate of a

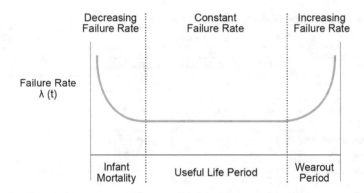

Fig. 3.10   Component failure rate over time for component population [25]

component population versus time, it is likely the graph would take the "bathtub" shape shown in Fig. 3.10. In this figure, the Y axis represents the failure rate and the X axis is time. From its shape, the curve can be divided into three distinct: infant mortality, useful life, and wear-out periods.

The initial infant mortality period of the bathtub curve is characterized by a high failure rate followed by a period of decreasing failure. Many of the failures associated with this region are linked to poor design, poor installation, or misapplication. The infant mortality period is followed by a nearly constant failure rate period known as useful life. The wear-out period is characterized by a rapid increasing failure rate with time. In most cases this period encompasses the normal distribution of design life failures.

Figure 3.11 shows the influence of effective building O&M on the performance of a building (and its components). This shows how a building (and its components) will eventually degrade in two scenarios, one with and one without "normal" maintenance. Of interest in the figure is the prolonged service life achieved through effective O&M. Not shown in this figure is the additional benefit of reduced building (energy) operating costs resulting from effectively maintaining mechanical and electrical equipment (e.g., lighting; HVAC; controls; and onsite generation).

The term "reactive maintenance" is commonly called a "run it till it breaks" maintenance mode. No actions or efforts are taken to maintain the equipment as the designer originally intended to ensure design life is reached. The evaluation of current practice shows that currently more than 55% of maintenance resources and activities of an average facility are reactive and about 30% are initiated on the basis of a maintenance plan.

M&R are not intended to alter or change the asset or to increase the useful life of the asset, but rather to sustain the asset in its present condition. M&R costs incurred to keep a fixed asset in normal operating condition will be expensed. Maintenance costs are not capitalized and are not recorded as part of the associated asset in the fixed asset record.

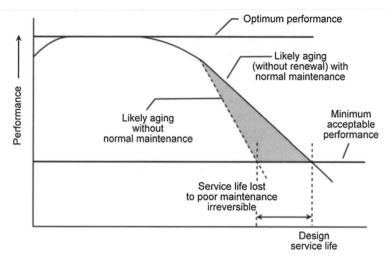

Fig. 3.11 Effect of adequate and timely maintenance and repairs on the service life of a building

On the other hand, **betterments** enhance an existing capitalized individual or group asset to a condition beyond that achieved through normal M&R. Betterment increases the useful life of the asset by at least 1 year without the introduction of a new unit. Only alterations that significantly rebuild an asset will be capitalized as betterments. The International Public Sector Accounting Standard (IPSAS) 16 "Investment Property" and IPSAS 17 differentiate between maintenance and "betterments" in the way that only betterments are eligible for the public accounting.

According to a study by the City of Houston and regulations of the Canadian Government [26], improvements or betterments of noncapitalized assets that do not involve replacements will be capitalized as part of the original asset only if the total cost of the original asset, including the improvement, is equal to or greater than $5,000. Otherwise, the improvement will be expensed as M&R.

Benefits from avoided M&R costs

Benefits in the M&R of the building after DER are mainly related to:

1. The reduced maintenance cost due to replaced, worn-out equipment at the end of its life cycle as a percentage of the new investment value, which would have been on the to do list and budget of the building´s owner and
2. The fact that "green" buildings cause less. RMI experience and studies on the correlation between green buildings and M&R costs indicate that green buildings generally have less M&R cost than the average building (in the range of 5–10%).

A 2008 Leonardo Academy study found that properties certified with LEED for Existing Buildings (LEED-EB) had a median M&R (not including janitorial) cost of $1.17 per square foot (12 €/m²) compared to the regional average of $1.52 per square foot (16 €/m²). After accounting for higher janitorial costs ($1.24 vs. $1.14

per square foot), the overall cost of maintenance was \$0.25 per square foot ($2.5 \,€/m^2$) cheaper, or a 9% annual maintenance cost savings. A study [27] conducted for the GSA found that 12 green GSA buildings had maintenance costs on average 13% less than the baseline.

A DER implements measures that replace worn out equipment, components, building parts, and applications that have previously been a routine part of the public building owner's life-cycle cost expenditures.

Usually these parts have achieved the end of their technical life cycle period and thus have a high demand for M&R and have a high default risk. This however does reduce availability of systems, energy security, and functionality of the building. In public buildings in many European countries a huge M&R backlog exists [28].

The cost saving potentials related to the maintenance costs is described in:

$$CSP = L_{Avoided} + MCR_{Avoided} \qquad (3.6)$$

CSP   Cost saving potential
$L_{Avoided}$ Avoided losses from default of functional units
$MCR_{Avoided}$ Avoided maintenance costs for the replaced equipment.

- The risks of default and reduced functionality or partially closed public buildings can only be estimated for certain types of public buildings:
- Hospitals, rehabilitation and recreation facilities, dormitories and staff buildings in the military context, where losses of functionality are usually expressed in lost revenues per bed and day. These data are available.
- Swimming pools, where lost revenues are recorded per day [29].
- Public IT server utilities.
- In all other cases, only the direct impact of avoided costs for the replacement of old equipment can be considered.

Baseline Building for M&R Savings

Cost data will be collected from the facility management system over a time period of at least 5 years. Figure 3.12 shows a typical M&R cost chart of a condensing boiler heating station (straight line). A review of the measures carried out over a 10-year time period show that a short time period would provide inconsistent (i.e., too low) cost data.

Calculation methods to determine the baseline

Public building agencies often cannot provide eligible maintenance cost data of their existing equipment so that a baseline can only be calculated.

(a) *Investment cost related M&R calculation*

In buildings with a large refurbishment backlog and the possibility of secondary damages, a simple estimate (see b) may not be sufficient to cover all the necessary measures to be carried out. In this case, the status, the damages, and the needs for

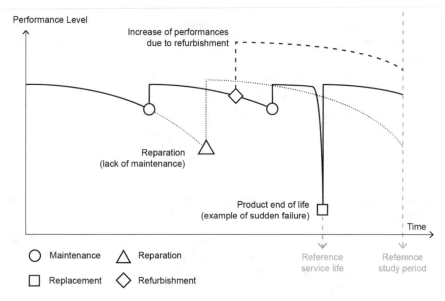

Fig. 3.12 Example of maintenance measures in relation to the performance over time (reference: EeBGuide, 2013, http://www.eebguide.eu/)

refurbishment must be evaluated in a schematic building audit. Condition assessments are an important aspect of effective maintenance planning. The incorporation of condition assessments as part of maintenance process programming ensures that there is a structured, objective process for identifying the demand for condition-based maintenance work to meet strategic and operational priorities. Such work should form part of any comprehensive program of maintenance in conjunction with preventive, statutory, and reactive (unplanned) maintenance work over the immediate, medium, and long terms.

These building condition assessments are common in the real estate sector. A number of templates exist, including:

- Australian Government: http://www.hpw.qld.gov.au/SiteCollectionDocuments/MMFBca.pdf
- Indian Ministry for Housing Affairs: (http://nidm.gov.in/PDF/safety/earthquake/link13.pdf)
- United States: https://www.usbr.gov/mp/berryessa/docs_forms/kleinfelder/chapter_02.pdf.

The condition assessment will lead to a directory of measures that will be evaluated with specific refurbishment costs, and to a first rough investment cost estimate for the works that:

- will have to be carried out in the next 1–2 years to eliminate the current M&R backlog

Table 3.7 Life-cycle period of ECM according to German Industry Standard VDI 2067.1

Appliance bundle (selected samples)	VDI 2067/B 1 (DE) (years)
Boiler/furnace	20
Air-Handling Unit (without distribution)	20
Cooling (water cooled rotary compressor)	18
Control (hybrid with pneumatic, electric and digital signal)	12
District heating pipes	40
Thermal envelope	40

- and will also determinate the annual costs to keep the building and its equipment in shape.

These investment costs and the net present value of future annual costs will be added and then distributed as annual M&R costs. The cost benefits of a DER can take into account the value partition of the measures carried out in the DER.

For example, the condition assessment sets up the following replacement investment:

- Building envelope: €200 k
- Boiler system: €100 k
- Considering a 20-year term and 2% interest rate, the annuity of the investment costs of these two measures is €18.5 k/year
- The annual costs for M&R are estimated by €20 k/year
- From the perspective of the building owner the annual baseline costs are k (20 + 18) €/year = k38€/year in addition to other part life cycle savings.

(b) *Related to the investment costs of new equipment—German Industry Standard VDI 2067*

According to the industry standards in some European countries (e.g., German Industry Standard VDI 2067, B1), maintenance costs are estimated as an average percentage value of the primary investment over the technical life time of this investment. In Belgium and the Netherlands [30] a similar calculation scheme is in use.
Example calculation:

- Primary investment of a boiler today: €50 k
- Life time: 20 years
- Germany: VDI 2067 (industrial standard) gives an average value for boiler (including installation) of 2%
- €50 k × 2% = €1 k/year maintenance costs in average over 20 years
- At the end of the technical life time period this value is 4% of the primary investment.

Tables 3.7 and 3.8 list related information pertaining to the technical life time of building components.

Table 3.8 Life cycle period of constructive measures according to German Industry Standard VDI 2067.1

Type of element	Position of the material (relative to the structural layer)	Location	Service life (years)	Example(s)
Roof	Structure	–	RSP	Concrete, rafters
Roof	External	Against exterior, flat roof	30	Insulation, waterproofing, vegetal layer, vapor barrier
Roof	External	Against exterior slanted roof	40	Tiles, lathing and counter-lathing, weatherproofing
Roof	External	Against ground	40	
Roof	Internal	–	40	Insulation, vapor barrier, coatings
Wall	External	Against ground	40	
Wall	External	With external insulation	30 15	Insulation, roughcast, boarding Paint, varnish
Wall	External	Without external insulation	40 15	Roughcast, boarding, Paint, varnish
Wall	Structure	Bearing or not	RSP	Concrete, bricks, wooden frame
Wall	Internal	–	30	Insulation, vapor barrier, coatings
Windows/door	–	Against exterior	20	
Floor	Internal		30 25 15	Hard coating: Ceramic tiles Medium Coating: Wooden or synthetic parquets Soft coating: Carpets
Floor	Internal	Between the structure and interior	30	Floating screed, water sealing, insulation
Floor	Structure	Above ground or cellar	RSP	Concrete, wooden beams

(continued)

Table 3.8 (continued)

Type of element	Position of the material (relative to the structural layer)	Location	Service life (years)	Example(s)
Floor	External	Above ground	RSP	Under floor insulation, light concrete, etc.
Floor	External	Against exterior	40	Insulation, coating

(c) *Swiss cost estimate for M&R (Meyer/Christen 1999)*

According to LCC assessments in the Swiss housing and commercial building sector, the annual M&R costs necessary to provide the building functionality are 1.0–1.5% of the present building insurance value. This can be considered as a M&R cost baseline if no other data exist.

Calculating post-DER M&R costs

To calculate the post-DER M&R costs, one must set up a relatively detailed M&R plan for the future that enables a calculation of future M&R spending. Relevant time boundaries are the technical lifetime of the system/equipment, after which the default risk of component(s) failure increases dramatically. Often smaller public building agencies and municipalities have never done this before.

The outcome of a M&R program is:

- The annual services for the maintenance and the controlling of the major components of the building
- Planned refurbishment (replacement) schedules with an indication of specific building component parts that must be replaced (e.g., the piston rod of a CHP engine after 30,000 h of operation).

Professional real estate managers, insurance companies, banks, and building auditors usually have national standards available that provide recommendations on how to set up M&R plans for the major components. Also equipment manufacturers offer mandatory M&R programs for their equipment; the technical and functional performance guarantee of 2 or 5 years is often combined with the execution of the M&R program.

Cost calculation of the M&R Program:

To calculate the costs, the M&R program will be tendered and skilled companies will be invited to participate. However, the functional performance guarantee is, in most cases, limited to 2 or 5 years.

Specific service companies, which in many cases are ESCOs, will be considered if the M&R program is combined with a functional performance guarantee of a

Table 3.9 Average M and R program costs for a time period T of 15–20 years in relation to the value of the first investment and per m^2 total net floor space in comparison to the average value of VDI 2067 B1 (Germany)

Component	T1 < 15 years	15 year < T2 < 20 years	VDI 2067 B1 of existing equipment at the end of the life cycle	Saving potential €/m^{2}year in comparison to existing equipment for T1 and T2
Natural gas boiler station	0.6–0.8% 0.1–0.15 €/m^2	1.4–1.8% 0.25–0.3 €/m^2	4% 0.45–0.5 €/m^2	0.25–0.35 €/m^{2}year 0.2 €/m^{2}year
Ventilation system	n.a.	2–3% 0.3–0.4 €/m^2	6% 0.6–0.75 €/m^2	T2: 0.3–0.4 €/m^{2}year
Building automation	1.5–2% 0.26–0.32 €/m^2	3–4% 0.4–0.52 €/m^2	8% 0.9–1.0 €/m^2	T2: 0.5–0.65 €/m^{2}year

component beyond 2 or 5 years. It is assumed that ESCOs have a strong sustainable interest in maintaining the functionality and performance in the best way.

The costs of M&R program will be determined by the responsibilities and services included; the major cost drivers are:

- Whether the program is more or less equal to the technical life time period of the equipment—as it is very likely that the component may have to be replaced over time
- Whether the full functionality is guaranteed over a time that is close to the technical life time period
- Whether the recovery of the full functionality in the case of default of the component must be guaranteed in a short time period.

An assessment of 28 ESCO projects in which the ESCOs provided a functional guarantee for all implemented components over a period of 15–20 years shows the following costs (Table 3.9). The relation is presented as the value of the first investment. Note that comparable values do not exist for thermal insulation.

Accountability of avoided maintenance costs

Avoided M&R costs can be quantified and measured where there is a requirement to calculate these costs. The benefit is clearly related to the added value of the replacement of a component. This allows for an accounting of calculated cost savings as described before.

The acceptance of accounting the avoided M&R is different from country to country and within the different public bodies. In Belgium and Germany, the majority of municipalities accept calculated approaches. On the level of German Federal government and in some of the Federal states, these costs benefits can only be accounted under certain framework conditions.

Value: According to the FM Market Report 2015, the avoided maintenance costs in DER projects can range from 20 to 40% of the energy cost savings in Germany.

3.7   Operation Cost Reduction by Building Automation Systems (BAS)

Operation costs are usually calculated at the points of operation, which are typically where building automation resides, and where the various centralized and detached equipment is installed in a building. Operation staffs usually have operation guidelines that contain to-do lists featuring the equipment and the necessary operations to consider at different time intervals. The goal of the operational guidelines is to ensure optimal availability, failure-free function, and a high level of performance.

In a DER, the technical equipment is often consolidated in combination with a refurbishment of the thermal envelope. To a certain extent, the operation is taken over by a modern BAS. The integration of more complex systems like heat pumps, CHP, and especially biomass heating will require a more intense operation process that increases the amount of onsite control that to a certain extend can only be delegated to a BAS.

The USDOE FEMP Program's O&M Guide [23] offers a number of metrics that can be used to evaluate an O&M program. Not all of these metrics can be used in all situations; however, a program should use of as many metrics as possible to better define deficiencies and, most importantly, to identify relevant potential improvements within a DER.

- **Capacity factor**—Relates actual plant or equipment operation to the full-capacity operation of the plant or equipment. This is a measure of actual operation compared to full-utilization operation
- **Work orders generated/closed out**—Tracking of work orders generated and completed (closed out) over time allows the manager to better understand workloads and better schedule staff (Fig. 3.13).

A modern building automation takes care of a variety of issues that may lead to quantifiable results:

- **Lower electricity and natural gas costs**—One of the major reasons that so many companies consider building automation is because a BAS allows the implementation of energy retro-commissioning to existing and newly installed HVAC equipment. The BAS collects energy performance data from global utility meters, and from a sub-metering structure that serves to collect consumption data from different building zones and high level single consumers. These data are transferred in the controlling level (automation level) and the management level. Direct control and adjustment will be carried out on the automation level (e.g., by using a heating load curve over the outdoor temperature) while trends and alarms will be provided

APOGEE BACnet System Architecture with Field Panel Web Server

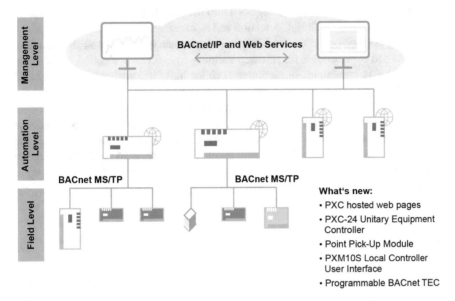

Fig. 3.13 BAS System and the three structure levels (reference: Siemens BT)

Table 3.10 Baseline building and estimated savings for operational costs

Operation activity	Not or only partly BAS based services	BAS based services	Savings (€ per m² year)
Energy management	0.16–0.2 €/m²year [34]	0.08–0.1 €/m²year [35]	0.08–0.2 €/m²year
Fault detection and clearance	>4 €/m²year	1.5–2.5 €/m²year	2.5 €/m²year

on the management level, e.g., by using an energy signature (kWh/day over average temperature per day). The evaluation of installed BAS shows that, on average, the savings can equate to about 1.0–1.5 € in electricity costs and 1.2–1.8 € saved per square meter of floor space. These savings can be achieved with a very small staff. By comparison with a non-BAS based energy management system, the BAS allows a reduction in operation hours for the energy management and controlling by at least 30% and up to 50%, which can account for up to 0.1–0.2 €/m²year (Table 3.10).

- **Maintained comfort levels**—A BAS can manage the comfort of both clients and employees in different usage zones with minimal staff effort. The assessment of more than 5,000 public buildings in Germany shows that most buildings have automatic heating, cooling, and ventilation controls. Since these control systems are not commonly integrated into a centralized BAS, it would be difficult to calculate any further operational staff cost reductions.

- **Simplified fault detection and clearance and improved maintenance management**: BAS systems that are integrated into the building default and maintenance management reduces default detection and replacement costs. IFMA reports show a saving potential of more than 25% in default management costs, not including the avoided expenditures for misused building space. The evaluation of data provided by FM companies in commercial office buildings with medium equipped HVAC and no specific backup requirements show costs of 1.5–2.5 €/m²year for BAS based services and >4 €/m²year for partly BAS based systems.
- Experience from ESCo projects that switch from an analog to a digital BAS without involving biomass show that 2 years after the refurbishment was accomplished and the BAS was adjusted to the building timetable on a detailed level, the operation workload and the time scheduled was shortened by at least 20% [31] and up to 40% [32].

3.8   Insurance Costs

One of the side effects of a DER is the replacement of outdated infrastructure. In public building administrations most of these risks are covered by building and hazard insurances. After the refurbishment, the contracts and premiums may be renegotiated and drastically reduced. In the context of this guideline, the premiums have been compared between non-refurbished and refurbished public office buildings of 5.000 m², with average equipment, including natural hazards, built in 1970–1980, and an insurance bundle including fire, ice and water damage, burst pipe, boiler and machinery insurance and, important for the IT infrastructure in public building including power shortage insurance.

With total costs of 21–30 €/m²year for the overall insurance package, the reduction of premiums is significant. The data in Table 3.11 show the scope of results (discounts) in comparison with a non-refurbished building (legend: ++ (30–40%), + [0–30%],—[no discount]).

Table 3.11   Reduced Insurance costs of DER buildings for an insurance bundle

Insurance risks DER measures	Fire and wind damage (35%)	Ice, water damage (15%)	Burst pipe insurance (24%)	Boiler and machinery insurance (12%)	Power failures (14%)
Windows	++	++			
Thermal envelope					
Duct and pipe systems	++	+	++	+	
Electrical system	–	–	–	+	++
HVAC	++	+	++	++	+

For an office building of 5,000 m^2 with insurance costs of 25 €/m^{2}year, with a new hot and cold water distribution grid is installed, the risk premium savings can be estimated as:

LCSP: 25 €/m^{2}year × Cost partition burst pipe 24% x savings 35% × 5.000 m^2 = 10.937 €/year.

3.9 Synergy of Investment and Maintenance Cost Savings

When delta-investment costs are considered, the delta-investment cost savings must also be assessed.

Often in appropriately funded projects, the public building agency schedules a DER process with limited investment cost budgets in several stages. By comparison with this "base case," a single-stage DER project carried out in collaboration with a deep renovation will be able to accomplish more measures.

Synergy effects:

1. Synergy effects will be derived from a better average U-value for the building envelope and the downsizing the heating and electricity load of the supply side. In the case of a 10,000 m^2 building with 65 W/m^2 heating load in the "base case" and 48 W/m^2 in the DER (new building) scenario, the heating load may be reduced by 170 kW, which is (with 55 €/kW for condense boilers from 500 to 750 kW) a synergy investment effect of 0.93 €/m^2 (which is less 0.5% of total delta investment costs of 180 €/m^2).
2. Avoided re-installation of building site equipment: in a one-stage process the multiple site equipment costs can be avoided.
3. Economies of scale and avoided annual inflation rate.
4. LCC savings will be realized in year 1 instead of increasing over time.
5. It can be assumed that a staged process will initiate less savings as if the DER is carried out in one stage. EuroPHit considers energy savings of 3–5% in a staged process.

   These potential synergies may be depicted in a cash flow analysis and a modeling calculation.

Synergy savings in a one-staged process:

For a first conservative estimate it can be assumed:

- **Investment cost savings**: In a one-stage process the investment costs can be reduced by 3–5% in comparison to a staged process.
- **Energy Savings**: In a one-stage process, the energy savings will be 5% higher than in a staged process.

3.10   Building Comfort

There is growing evidence from around the world showing non-energy-related economic implications of sustainable renovation of buildings, specifically DER. In these studies, benefits resulting from DER are linked either to different green building certification systems, e.g., building certification system LEED® (USA), BREEAM (UK, EU, EFTA member states, EU candidates, and the Persian Gulf), Green Star (Australia), CASBEE (Japan), DGNB in Germany, or to energy-only focused programs, e.g., EnergyStar or Passive House Standard. Obviously, transparent green building certifications provide a basis for investors to measure and compare properties, a critical foundation for financial analysis [35]. While Energy Star and the Passive House certificates are solely focused on energy, different green buildings certification programs, in addition to energy and water saving attributes, include emissions, waste, toxicity, and overall environmental performance criteria designed to reduce the overall impact of the built environment on human health and the natural environment. Many of these sustainable features considered in different certification systems are directly related and required for successful DER project.

Many sustainable features have multiple impacts on the property value. For example, daylighting can contribute to worker productivity and thereby increase rents. It can also reduce energy costs and thereby reduce operating expenses. Daylighting, if not property implemented, can also result in glare and/or thermal comfort problems. External wall insulation and high performance windows reduce cold or heat radiation and therefore make people more comfortable at workplaces located along the perimeter. A DOAS with well-designed air distribution effectively provides sufficient outdoor flow rate, prevents cold drafts and improves thermal comfort at work stations. The GSA has reported [36] the following benefits resulted from best practices of building renovation projects:

- **Reduced absenteeism**: Healthier indoor environments reduce sick building symptoms and absenteeism. A Canadian study revealed that approximately one third of employees' sick leave can be attributed to symptoms caused by poor indoor air quality. The same study found that communication and social support enabled by open office plans are strong contributors to healthy workplaces and lowered absenteeism. According to a study by Carnegie Mellon University (CMU) for the USDOE, improving indoor air quality and providing natural light reduces illness and stress. The CMU study [4] showed that occupants closer to windows reported fewer health problems. In addition, a survey of three case studies by the Rocky Mountain Institute proved that better lighting and HVAC systems could reduce absenteeism from 15 to 25% [37].
- **Increased productivity and performance**: Flexible, adaptable work settings allow people to customize their workspace to suit their individual needs, providing improved comfort. When given control over their environment, workers are less distracted and more productive and satisfied with their jobs. They also report fewer complaints to building management. For example, Public Works and Government Services Canada found that when people were given individual ventilation control,

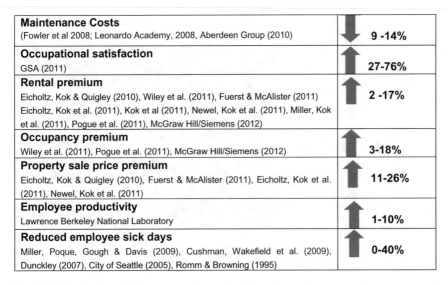

Maintenance Costs (Fowler et al 2008; Leonardo Academy, 2008, Aberdeen Group (2010))	9 -14%
Occupational satisfaction GSA (2011)	27-76%
Rental premium Eicholtz, Kok & Quigley (2010), Wiley et al. (2011), Fuerst & McAlister (2011) Eicholtz, Kok et al. (2011), Kok et al (2011), Newel, Kok et al. (2011), Miller, Kok et al. (2011), Pogue et al. (2011), McGraw Hill/Siemens (2012)	2 -17%
Occupancy premium Wiley et al. (2011), Pogue et al. (2011), McGraw Hill/Siemens (2012)	3-18%
Property sale price premium Eicholtz, Kok & Quigley (2010), Fuerst & McAlister (2011), Eicholtz, Kok et al. (2011), Newel, Kok et al. (2011)	11-26%
Employee productivity Lawrence Berkeley National Laboratory	1-10%
Reduced employee sick days Miller, Poque, Gough & Davis (2009), Cushman, Wakefield et al. (2009), Dunckley (2007), City of Seattle (2005), Romm & Browning (1995)	0-40%

Fig. 3.14 The direct and indirect cost savings of a DER estimated based on industry reports and studies (RMI 2015b (111))

the number of trouble calls decreased significantly [K. E. Charles, et al. 2004]. Figure 3.14 estimates the positive impact that some of these direct and indirect values beyond energy cost savings can bring as the result of a DER.

In Belgium, the comfort meter EPC business model has been implemented in a public bank. This project captures, besides energy savings, two additional bankable values:

- Building performance and comfort parameters. If an ESCo can operate the building in a comfort zone, a bonus-malus related payment will generate revenues for the ESCo. The revenues are related to the result of an analysis of the cost savings, which can be achieved from increased productivity and reduced absenteeism within defined comfort performance parameters. The maximum remuneration that can be achieved by the ESCo from optimized comfort parameters is 10 €/m²year.
- Added asset value: at the end of the contract period, a bonus-malus payment according to the increased or declined elements and building value is settled. After the end of the ESCo contract period, a team of building auditors will assess the building and the actual condition of the building's technical equipment. If, at the end of the contract period, the value of the equipment is still >0€, an additional value is achieved, e.g., by appropriate maintenance. The ESCo is awarded with a payment that equates to 50% of the additional value at the end of the contract period (Fig. 3.15).

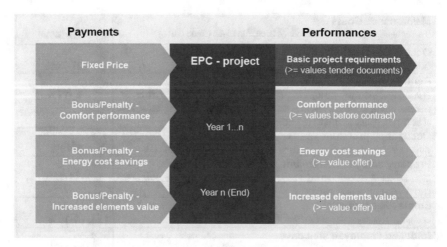

Fig. 3.15 Revenues within the EPC project. *Source* Factor4

3.11 Impact of DER on Building Values

The option of accounting the added value provided by DER by adopting the evaluation methodologies of the real estate evaluation could contribute to the cost-effectiveness of DER projects. The evaluation of public real estates is comparable to the commercial sector.

1. Location of the building
   The location plays an important role in the evaluation of buildings. The assessment of several different real estate evaluation guidelines shows that the DER project evaluation is usually carried out using matrixes that consider the following criteria (described in their extrema in Table 3.12):

   a. Building density allowance: In urban planning processes in large cities the density allowance factor becomes more important. To provide affordable living and working space areas, the density allowance factor is often increased. This provides the option of increasing the available floor space without consuming additional land. DER can create additional floor space by turning cold roofs into usable warm roof areas, which can be combined with additional mansards.

   b. Attractiveness of the area: Public buildings often use their buildings to start the "re-socialization" of a worn down area. A DER project integrated in a major building renovation may be the starting point for increasing the attractiveness of an area. Since this value-added impact for the area is only considered over a term of 5 years or more, it would be difficult to quantify.

Table 3.12 Evaluation of office building location values

Criterion	Low value	High value
Size and type of the settlement	Remote, outlying area of small villages, holiday resorts with no specific highlights	Mid-size large size cities, inner city, shopping area, exclusive holiday resort
Commercial or dwelling area	Industrial production area	Office neighborhood close to dwelling area
Building density allowance	<30%	70% < BDA < 100%
Attractiveness of the area	Low frequented, declining and worn down building and critical social structure	High frequented, upgraded or new building structure, well working social structure
Emissions in the area	Highly polluted	Low emission area, clean air plan enforced
Connection to public transportation, street infrastructure, internet, WLAN and commercial retail and other infrastructure	Not integrated	LWC internet, good to excellent connection to public transportation, good accessibility to individual traffic
Unspecific		Gentrification process in maturity phase

c. Emissions: A DER can reduce local emissions as it reduces the site energy demand and considers supply solutions based on renewable energies. The quantification of the added asset value has not yet been accurately assessed.

2. Real Values and DER:
   Evaluating the substance and the actual market value also considers the replacement costs of misused building areas and equipment. The replacement costs are calculated in two separate calculations by two main factors: constructive value and the land value. The real values need not necessarily consider revenues. This method can be valuable for public buildings without rental incomes.

   a. The constructive value is calculated using the current construction and specific prices for the building's reconstruction. In the second step, the devaluation of the building is estimated using its current status, along with the labor and material costs required to upgrade the building to serve its current purpose.
   b. The land evaluation uses criteria (referred to in Table 3.12) and cost values related to each criterion on a regional level e.g., a congested urban area.

c. Example:

1,200 m² floor space x 2.800 €/m² new construction costs	= € 3.36 M
Devaluation:	
Replace rendering: 3,000 m² surface x 60 €/m²	− € 0.18 M
Install new elevator	− € 0.40 M
Replace windows: 1,000 m² surface x 490 €/m²	− € 0.49 M
Net construction value	= € 2.29 M

An evaluation of the way DER measures help to increase the value of the real estate must consider two criteria: why green colors in the figure above?

d. **Is the value accountable under public accountancy regulations?** A DER investment must consider the cost effectiveness of the expenditures (investment costs and annual operative costs) and its accountability. "Accountability" means that the monetary value of the investment is, at least partially transferred to the accountancy and balance sheet as a positive value (building asset). The market will then decide, e.g., by considering the land value, how much of the added value provided by a DER project may be relevant to the market price. The added value will be more relevant in high level buildings. By no means can it be said that an added value will in practice be fully accounted in its entire value.

e. **Accountability**: Accountancy regulations are different from country to country for non-stock market building owners; stock market or municipal bond based companies need to consider the IAS, IFRS, and to some extent, the U.S. GAAP accountancy regulations [38–40]. Non-stock market building owners will have to follow national accountancy regulations. The following section describes a number of best-practice examples in the public and commercial sector that show pathways to account for added values to the accountancy sheet.

In some countries, the public accountancy regulations allow for an accounting of the added value of DER at least to the level in which DER measures create "**a betterment**" of the situation that existed before the DER project. This might include, for example, the installation of thermal insulation on a plastered wall, which would be a "betterment"; the plastering itself would not be considered a "betterment." In a few cases, the value of the betterment is accounted at 80–100% of its value at the day of implementation (usually the purchase price). However, that value quickly depreciates.

3.11.1 International Regulation Framework

However, the option to account the added value must be proved in every public sector in each country individually; the following samples provide some first indications.

- **Commercial buildings in Germany**: In Germany, the Commercial Code HGB
  Sect. 255, Chap. 2, s1 is the legal reference for the activation and entry on the assets
  side. The option for the activation of DER measures allows that the DER measures
  become an economic good, which provides the option of tax (of earnings) reliefs,
  write offs. The following expenditures can be activated in the balance sheet of the
  commercial building owner:

  - Increasing the floor space must be considered from the perspective of the invest-
    ment costs and operative expenditures, but also from the perspective of addi-
    tional usable commercial space.
  - New implementation of thermal envelope, HVAC, and supply solutions that
    have not been in place before.
  - Major improvements: If, for example, double glazed window were already in
    place pre-DER, the additional costs for the triple glazed window and additional
    functionalities may be accounted.
  - However, the plastering, which may have been in place before cannot be consid-
    ered. Common costs such as scaffolds may be considered to the part as they are
    related to the first implementation or to the major improvement. For example,
    if the plastering and the insulation measures relate 10:90 in terms of investment
    cost values, 90% of the scaffold may be considered for activation as well.
  - It is required to present building value assessments at least every 5 years.
  - From the perspective of the market value, the DER measures have been, at least
    to the allowable level been put into the balance sheet of the building. That means
    that, in the case of a buildings purchase, these costs will have to be a part of the
    purchase price. This is the same with the calculation of rental rates.
  - For public buildings using the International Public Sector Accounting Stan-
    dards: International Public Sector Accounting Standard (IPSAS) 16 "Invest-
    ment Property" and IPSAS 17 provide the framework for the public sector. Also
    a differentiation between maintenance and "betterments" is made here; only
    betterments are eligible for the public accounting.

- Canada: Guidance for Local Governments and Local Government Entities:

  - The cost of an asset will also include subsequent expenditures for "betterments."
  - The "Betterment" is a cost incurred to enhance the service potential of the asset.
  - In general, for tangible capital assets, service potential is enhanced; this is the
    case when there is an increase in the previously assessed physical output or
    service capacity, associated operating costs are lowered, the useful life of the
    property is extended, or the quality of the output is improved.
  - Any other expenditure not fulfilling these criteria would be considered a *repair
    or maintenance* and expensed in the period without being considered as an
    asset in the accountancy sheet of the public entity.
  - For complex, long-lived network systems, it is more difficult to distinguish
    between maintenance and betterment. It is not always practical to determine
    whether an expenditure will or will not extend an asset's useful life. *Maintenance
    and repairs* maintain the predetermined service potential of a tangible capital

asset for a given useful life. Such expenditures are charged in the accounting period in which they are made. Betterments increase service potential (and may or may not increase the remaining useful life of the tangible capital asset). Such expenditures would be included in the cost of the related asset.

3.11.2   Income Related Real Estate Evaluation: Gross Rental Method (GRM) and Discounted Cash-Flow Method

The income related evaluation methods only take into account the revenue streams of an asset. The GRM considers the net present value of annual revenue streams of rental rates or comparable other income over a defined period of time. GRM must consider the impairment of rental incomes with increasing building depreciation.

The Discounted Cash Flow Method is comparable to the NPV and considers expenditures and revenues (annual cash flow), increasing asset values, and depreciation over a time period of 15–30 years. The method is considered for public buildings with an income such as rental rates, usage fees, or other money transfers from building user to building owner.

3.11.3   DER and Green Building Values in Public and Non-public Sectors

Significant progress has been made in the real estate industry in quantifying and articulating the value of sustainable property investment. Even though most investors, and many tenants, understand that sustainable properties can generate health and productivity benefits and recruiting and retention advantages, and can reduce risks, they struggle to integrate benefits beyond cost savings into their valuations and underwriting.

However, this concept has not yet found a broader traction in the public sector. As the result, most of decisions related to the scope of energy and sustainability work under building major renovation project are based on cost effectiveness resulted from operating cost reduction.

To promote the concept of DER value to the public sector, consider a trend of building value change throughout its useful life using schematic in Fig. 3.16. Assume that the building has some value **V1** "on the books" upon construction completion, and at the beginning of its operation. Throughout its operation, the building value depreciates to the point where it has some residual value **V2** at the end of its useful life, when major renovation is warranted. Throughout the process, the value can slightly increase with some minor renovations or drop due to manmade or natural damage to the building, which may require that major renovation is required before the planned end of its useful life. In a major renovation, the value of the building will increase

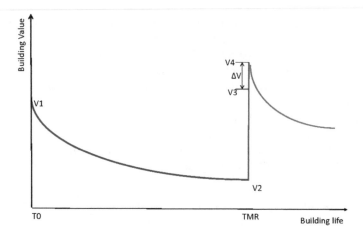

Fig. 3.16 Building value over its life

and reach the **V3**, which will probably be higher than **V1** due to more stringent standard building requirements (seismic, thermal comfort, indoor air quality, etc.) compared to those used during its design and construction. If the major renovation is combined with DER and follows energy and sustainability requirements beyond the minimum standard requirements at the time of its renovation, the value of the building increases to **V4** by $\Delta \mathbf{V} = \mathbf{V4} - \mathbf{V3}$. Based on the data available from the private sector studies described above, it is safe to suggest that $\Delta \mathbf{V}$ may be at least 5%, which, when annualized, can be used in an LCCA in addition to operating cost savings to justify the cost effectiveness of a DER.

A couple of studies [41] have been evaluating the economic advantages of energy efficient/sustainable buildings in the residential and commercial building sector in different regions. The evaluation focused on KPI relevant for the evaluation of the economic value of energy efficient buildings such as rental rate increase, time to sell, sales price, and consumer preference. So far, comparable data are not available for the public sector. However, these information can provide a first indication for comparable value impacts in the public sector. Analysis of 10,000 buildings in the United States labeled as LEED and/or EnergyStar compliant [42, 43] shows that otherwise identical commercial buildings with an Energy Star certification will rent and sell for considerably higher values. Together with data from other real estate markets, the impact of DER and Green buildings shows the following results in both the residential and non-residential sectors:

(a) **Residential sector (Studies comprise 2008–2016)**:
    The impact of energy efficiency standards beyond the national minimum require-
    ments on the sales price shows an impact of:

- 3.5–7% in Switzerland for a Minergie label
- 1.4€ per reduced kWh/m²year in Germany
- 5.5–9.5% in Australia for Efficiency Star ratings of +5

- 2.9% in the Netherlands for Energy Performance Certificates A, B, C
- Price reduction of 6–11% for Tokyo Green Labeled buildings in Japan
- 8.9% in the United States for LEED certified buildings
- A better marketability expressed in a reduction of the time to sell by 18 days in the United States for Built Green, Energy Star, Earth Advantage, or LEED certified buildings.

  The impact on the rental rate has been
- 6% in Switzerland for the Minergie label
- 0.38–0.5 €/m² (roughly 5–10%) in Darmstadt, Germany for source energy level below 250–175 kWh/m²year.

(b) **Commercial Sector (2008–2016)**

The impact of energy efficiency standards beyond the national minimum requirements on the sales price shows an impact of

- 30% in the United States for LEED certified buildings.

  The impact on the rental rate has been
- 6–17% in the United States for LEED certified buildings
- A recent Leonardo Academy study [44] that evaluated data collected during 2006–2007 from owners or managers of 23 LEED-EB certified building showed that, in all the categories of operating costs, more than 50% of the LEED-EB buildings had expenses 13% less than the average for the region ($62/m²year).

3.12   Increasing Available Floor Space

Many studies have also reported an increase in leasable space as a result of downsizing mechanical systems as part of a DER. For example, deep retrofit of the Deutsche Bank Twin Towers in Frankfurt, Germany, which freed up an entire building story to be used as newly leasable space. An important additional benefit resulting from installation of modern pitched roof insulation systems and the creation of thermally controlled attics is the increase of valuable usable space that can contribute to the cost effectiveness of the DER project. Figures 3.17 and 3.18 show examples of such roof renovations. The Technical Guide gives further technical details with best practices of roof insulation.

The hotel building was renovated in 2010. The roof and walls were insulated and high performance windows were installed in walls and in the roof. That increased the number of rentable rooms to 15. Windows now provide sufficient daylighting during the day and are equipped with shades to be used during nights (when there is little dark.)

The building was built between 1866 and 1872 and is now used as a combination of administration and technological shop building. The building had uninsulated roof and walls and single-pane windows; the attic was never used. In 2006, the building was upgraded to become office space. As the a part of the renovation, walls

Fig. 3.17 Example of attic renovation at the Klosterhagen Hotel in Bergen

were insulated with blow-in cellulose insulation, windows were replaced and gaps between windows and walls were sealed, the roof was insulated using spray foam insulation, and a new reflecting ambient lighting system was installed, and new air heating and ventilation systems were installed to meet indoor air quality (IAQ) and thermal comfort needs.

3.13 Conclusions: Life Cycle Cost Analysis and Optimization

Besides energy cost savings from demand-side reduction the integration of bankable part LCCs can improve the cost effectiveness of DER project significantly. For comparison, Fig. 3.19 shows the annual LCCs of three different LCC approaches, and of a German public office building, 3600 m^2, ASHRAE c.z. 5, with an EUI of 180 kWh/m^2year:

1. With a time period of 20 years, an interest rate of 3% over 20 years, the annuities of DER global investment costs for this building are between 28 €/m^2year, which

Fig. 3.18 Example of attic renovation of the building at Rock Island Arsenal. *Source* Library of Congress (a, b); Alexander Zhivov (c–f)

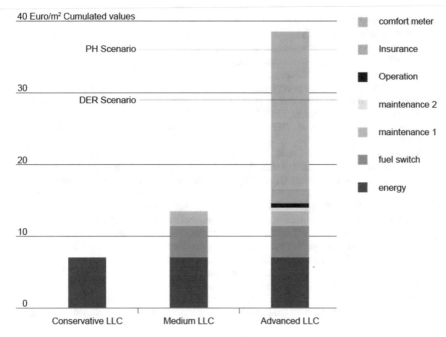

Fig. 3.19 Advanced LCC—cumulated values EUR/m^2 for three LCC scenarios

equates to a scenario "–51% against energy baseline" and 37€/m²year for global investment costs of a passive house scenario.

2. By integrating the maintenance and fuel switching to the energy savings, the total benefits increase by factor of 2.
3. By considering insurance, operation, and the comfort meter approach, even a passive house investment costs payback period is <20 years.
4. The least-cost planning calculation is necessary to calibrate and optimize the ratio of investment costs for demand-side reduction, energy supply, and the impact on the LCC.

References

1. References on national level: SIA D0213 (Switzerland), ZIA 100 Guideline (Germany)
2. BIMA Report 2012, Berlin, German
3. W. Pan, A.R.J. Dainty, A.G.F. Gibb (2012) Establishing and weighting decision criteria for building system selection ASCE. J. Constr. Eng. Manag
4. Prediction and optimization of life-cycle costs in early design, Bogenstätter http://dx.doi.org/10.1080/096132100418528
5. www.iso.org/iso; https://www.beuth.de › Startseite › Bauwesen › DIN EN 15804, eplca.jrc.ec.europa.eu/uploads/ILCD-Recommendation-of-methods-for-LCIA-def.pdf; VDI 2067, Blatt 1, Wirtschaftlichkeitsrechnung (Beuth Verlag, Berlin, 2010)

6. ISO/DIS 15686-5
7. U.S.A., Federal Energy Management Program rules (10 CPR 436) that the study period can be as long as 25 years, in *European countries the*
8. https://www.wbdg.org/resources/life-cycle-cost-analysis-lcca
9. For examples, see: https://www.ons.gov.uk/economy/nationalaccounts/balanceofpayments/datasets/tradeingoodsmretsallbopeu2013
10. Bleyl et al., Building deep energy retrofit: using dynamic cash flow analysis and multiple benefits to convince investors, EEECB paper draft (2017)
11. Moerck, Zhivov, Lohse et al., DER Case Studies. IEA EBC Annex 61, 2017 (print in preparation)
12. M. Hyland, R.C. Lyons, S. Lyons *The Value of Domestic building Energy Efficiency-Evidence from Ireland, 2012* (Department of Economics Discussion Papers, University of Oxford, 2012)
13. International Energy Agency (IEA), Capturing the Multiple Benefits of Energy Efficiency (Paris 2014)
14. Developing EPC towards DER, Lohse, ASHRAE technical paper (2014)
15. Efficiency Valuation Organization (EVO): International Performance Measurement and Verification Protocol, Toronto, 2015 (https://evo-world.org) The method outlined in IPMVP Option C (Whole Facility), supported by the data collection methodology described in EN 16247-2 Energy Audits –Part 2: Buildings [2c] should be followed
16. EPC Best Practice, Do Environment and Energy, Stuttgart, 2015 (German) http://www.energiekompetenz-bw.de/contracting/beispiele/gute-beispiele/
17. Efficiency Valuation Organization (EVO): International Performance Measurement and Verification Protocol, Toronto, 2015 Grid stabilization is adding approx.25% of the power production value
18. FGW Regulation No. 3, Germany; http://www.wind-fgw.de/pdf/TR3_Rev20-e_preview.pdf
19. Reversing the Deterioration of our Nation´s Public School Buildings (Council of the Great Cities, Chicago, 2014)
20. INQA Bauen "Instandhaltung von Bauten und technischer Gebäudeausrüstung http://www.inqa-bauen.de/upload/komko/pdf/7_3_0106.pdf
21. PECI. 1999. Operations and Maintenance Assessments. Portland Energy Conservation, Inc. Published by U.S. Environmental Protection Agency and U.S. Department of Energy, Washington, D.C (1999)
22. IFMA Members ID Top FM Best Practices and Drivers of Quality FM http://www.fmlink.com/Marketplace/WhitePapers/Articles/old/MACTEC-070907.html
23. O&M Guide, US Department of Energy, Washington (2012)
24. NASA O&M Guidebook (2009), https://nodis3.gsfc.nasa.gov/npg_img/N_PR_8831…_/N_PR_8831_002E_.pdf
25. O&M Guide, U.S. Department of Energy (2010)
26. Fixed Asset Accounting and Management Procedures Manual Section 10, Improvements, Betterments, and Maintenance.; City of Houston. Revision 4. February 3, 2005, http://www.houstontx.gov/finance/fixed_asset/fixed_asset_10.pdf "Guidance for Local Governments and Local Government Entities"
27. K.M. Fowler, E.M. Rauch, *Assessing Green Building Performance, A Post-Occupancy Evaluation of Twelve GSA Buildings* (Pacific Northwest National Laboratory, 2008)
28. KEA Reports: Contracting als Modell zur Erschließung von Energieeffizienzinvestitionen, Karlsruhe 2008 (German) Cost estimates show that to tradeoff the current refurbishment backlog in Baden-Württemberg´s public buildings would demand for a €15bn investment in the next 5 years
29. KEA Reports: Contracting als Modell zur Erschließung von Energieeffizienzinvestitionen, Karlsruhe 2008 (German) However most of the public swimming pools are not operated cost effective so that the calculation of losses might turn into a revenue calculation of avoided losses in operational costs
30. Dutch standard for condition assessment of buildings, Ad Straub, OTB Research Institute for Housing, Urban and Mobility Studies (Delft University of Technology, The Netherlands, 2012)

31. Kompetenzzentrum Contracting, Projektberichte, EPC projects in Ehingen/Donau and Oftersheim, Karlsruhe, 2016 (German)
32. Kompetenzzentrum Contracting, Projek, EPC projects in Buchen, Besigheim, Reutlingen, Freiburg Pool 2,3, Karlsruhe, 2017 (German)
33. Municipal Administration of Frankfurt/Main. Energiebericht 2009, Frankfurt/Germany: "9 Energy Managers for 250 buildings assumed with 15.000 m^2 average floor space"
34. According to time schedules of professional energy managers at ESCos supervising 50 and more buildings per capita in EPC projects (remuneration is related to the energy performance)
35. S.R. Muldavin, Value Beyond Cost Savings: How to Underwrite Sustainable Properties (2010)
36. Value Beyond Energy, GSA (Washington, 2015)
37. J.J. Romm, W.D. Browning, greening the building and the bottom line: increasing productivity through energy efficient design (Snowmass, CO: Rocky Mountain Institute; 1994), http://www. greenbiz.com/files/document/O16F8527.pdf. Accessed 10 May 20081998
38. International Accountancy Standards IAS Nr. 2, 5,11,16,36, 40 and US GAAP, International Financial Reporting Standards IFRS 5 and national Standards
39. S. Sayce, A. Sundberg, B. Clements, *Is Sustainability Reflected in Commercial Property Prices: An Analysis of The Evidence Base* (RICS Research, London, 2010)
40. Bio Intelligence Service, Ronan Lyons and IEEP, 2013, Energy performance certificates in buildings and their impact on transaction prices and rents in selected EU countries, Final report prepared for European Commission (DG Energy)
41. WGBC, 2013, The Business Case for Green Building - A Review of the Costs and Benefits for Developers, Investors and Occupants, World Green Building Council
42. P. Eichholtz, N. Kok, J. Quigley Doing well by doing good? Green office buildings. Am. Econ. Rev. 100 (December 2010): 2492–2509 (2010), http://www.aeaweb.org/articles.php?doi=10. 1257/aer.100.5.2492
43. S. Wachter, Valuing energy efficient buildings. report prepared for energy efficient building hub, http://gislab.wharton.upenn.edu/Papers/Valuing%20Energy%20Efficient%20Buildings.pdf]
44. The Economics of LEED for Existing Buildings (Leonardo Academy Inc., April 21, 2008)

Chapter 4
Funding of DER Projects—Financial Instruments

As soon as the cash flow plan is prepared, the financing of the project can be considered. Financing a DER requires cost-effective financing of high investment costs. For public entities, the impact of financing instruments on the balance sheet is relevant for decision-making. In recent years, the bank loans and soft loans have been complemented with a number of attractive financing instruments for the public sector. This chapter describes financing instruments and their usefulness for DER projects from the perspective of public administrators and financiers by the following criteria: flexibility, impact on the public debt balance, risk mitigation, cost effectiveness, and options for combining those instruments with other financing tools.

Financing DER-Perspective of the Financiers
In light of the banking crisis of 2008, the EU Capital Requirement Regulation and Directive (CRR/CRD IV) was formulated to apply to credit institutions and investment firms that fall within the scope of the Markets in Financial Instruments Directive (MiFID). Specifically, the risk-weighting under Pillar 1 of the CRD IV requires that the regulatory capital and the liquidity requirements (Liquidity Coverage Requirement Delegated Act) required for any specific asset be in line with the actual risk profile of that asset. The new regulatory capital requirements of Basle III impact EU banks put pressure on the availability of risk capital and on the balance sheets of all financial institutions, and impact energy efficiency investments in all categories. Concern is rising that these new regulations will be blind to environmental targets, and to the long-tail impacts of climate change and the stranded assets that unsustainable and low resilience investing can create in this context. Obviously, the required capital adequacy ratios may be inappropriate for Energy Efficiency investments. The accounting regulations for energy efficiency investments neglect to consider the value of inherent multiple benefits, which makes it difficult for financial institutions to allocate investment capital.

© The Author(s), under exclusive license to Springer Nature Switzerland AG 2019   77
R. Lohse and A. Zhivov, *Deep Energy Retrofit Guide for Public Buildings*,
SpringerBriefs in Applied Sciences and Technology,
https://doi.org/10.1007/978-3-030-14922-2_4

Fig. 4.1 Types of financial
instruments supporting the
energy performance of
buildings

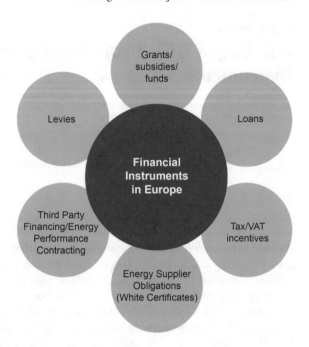

4.1 Financing Instruments

Conventional financial instruments that have been used since the oil crises of the
1970s include: grants and subsidies, loans, and tax incentives. Financing is also provided in international funds, either through European institutions such as "European
Bank for Reconstruction and Development" and the European Investment Bank, and
institutional (mostly environmental or green) funds. The formats used are mostly
soft loans and grants distributed by commercial banking institutions. The innovative
instruments include EPC (often known as Third Party Financing) and Energy Supplier Obligations (often known as White Certificates). The following definitions of
financial instruments and their function must be considered:

- Subsidies are handed out to reduce the investment costs of equipment and installations over a certain period of time, i.e., broadening the market approach of a
  quasi-mature product.
- Grants are targeted at the end consumers such as households, industrial entities, etc.
  to pay for a part of the incremental costs of introducing energy efficient processes
  in the market—such as enhanced building insulation.
- Grants or subsidies may be financed directly through the state or local authority
  budget, or through hypothecated taxes (also known as ring-fenced or ear-marked
  tax) (Fig. 4.1).

This following chapter briefly presents financing instruments with regard to the
specific needs of a public building owner in a DER project in the public building stock.

The specific demand of a DER financing instrument is related to the characteristics of the technical execution of DER projects. DER projects take place in the building stock and are typically carried out at a point of time more or less close to the end of the first life cycle period of the building. The time between the first construction of the building and the decision to carry out a DER is commonly at least 20 years or more, during which time the building undergoes numerous more or less well-documented adjustments, repairs, new installations, partly replaced infrastructure, etc.

Technical design standards, security standards, and stricter regulations on hazardous materials in the building have been changing over time. This creates certain risks that mainly result from inadequate information regarding a building's technical status, which impact the steps of design, implementation, and operation. From the perspective of the implementation of investment, these uncertainties create mainly risks that impinge on the DER project's time schedule and on the total investment cost. A handful of criteria must be considered to evaluate the DER financing instruments in the public sector:

- Flexibility of time schedule for the payment of the credit amount to the debtor.
- Flexibility to increase (or decrease) investment cost totals and expand the borrowed amount of money.
- Relevance for the debt-balance of the public building owner. Countries with a relatively strict austerity policy for the public budget law will have to consider how the financing tool will appear in the public accounting system—as a liability or debt. The assumption is that, in most cases, the credit worthiness of a public entity will, to a certain degree, be related to the same criteria that determine the investment grade of public or in the United States and the UK, public and municipal bonds [1]. In other European countries, the re-funding of the public sector is usually provided by special public loan programs such as the "Kommunalkredit" in Germany. The public sector is supervised by public finance and debt control systems, which are located in the ministries of interior (municipalities), the Federal ministry of finance (Federal sector), and which refer to the annual public cash flow budget that each public entity must provide. In this assessment, the creditworthiness is related to the balance of public income (taxes, etc.) and to the liabilities and obligations. Under these premises, loans for public entities are provided as "blank" credits, without securities.
- Inherited risks and the specific risks of DER projects. Which are the risks of a financial instrument, where are these risks allocated, and which mechanisms does the instrument provide to reduce the risks? Long term DER financing agreements bear certain risks such as increasing interest rates, and prepayment penalties that can have a significant impact on the NPV of the project. Also, the performance risk of the DER project must be considered here. From the perspective of the creditor, the normal assessment of the creditworthiness of the obligor cannot be predicted over a term of more than 15 years. The previously mentioned risks of delays and shifting investment cost total are also considered here.
- Cost effectiveness of the financial instrument. The cost effectiveness of a financial tool is related to the transaction costs necessary to prepare the financing agreement.

In many financing instruments, the transaction costs of DER projects are still above-average as the number and size of such projects are still comparably small and the due diligence process cannot refer to numerous well evaluated reference projects.

• Combination with soft loan and grant programs. Many EU countries provide a set of subsidiary tools to encourage DERs directly or indirectly by reducing first and annual costs of the DER investment. Well known programs such as KfW in Germany help to reduce interest rates and the repayment for refurbishment, which target more ambitious energy efficiency levels than the minimum energy requirements defined in the national standards. Tools that allow the combination with soft loans and grant programs provide a significant benefit for the public building owner.

4.1.1 Loan Financing, Credit Lines, Revolving Funds, Preferential Loans

A conventional bank loan is the most common form of DER debt in the public sector. After the cost estimation of the architects, the financing plan is set up and the demand for external funding is defined using the construction time schedule (Table 4.1). The bank loan is an agreement to lend a principal sum for a fixed period of time, to be paid back within a defined term; the interest rate is calculated as a percentage of the principal sum per year and other transaction costs. Soft loan programs are disbursed by financial intermediaries such as commercial banks. The loan structure depends on the obligor/creditor and on the type of measures to be financed. Loan terms may vary from 5 to 20 years. Typically, the interest rate will be fixed over a certain period of time and will be capped to a maximum throughout this time period. This allows a reduction of risks and opportunities from the interest rate level. The most common method is **annuity repayment** in which the interest plus principal repayment are a constant value over time.

4.1.2 Soft Loans/Dedicated Credit Lines

Soft loans are subsidized loan programs with no interest or a below-market rate of interest, or loans made by multinational development banks and government agencies to developing countries that would be unable to borrow at the market rate. Soft loans have lenient terms, such as extended grace periods in which only interest or service charges are due, and interest holidays. Soft loans typically offer longer amortization schedules (in some cases up to 50 years) and lower interest rates than conventional bank loans. A dedicated credit line provides low-interest loans to reduce capital costs.

Table 4.1   Assessment of eligibility for DER financing

Criterion	Eligible for DER financing?
Flexibility: The flexibility of normal bank loans is limited: The loan amount is fixed in the loan agreement. An increase of the loan amount would mean to set up a new loan agreement; alternatively a "floor limit" (maximum loan amount) or a credit line can be settled using a first detailed investment cost calculation and includes, if possible a cushion for contingencies. However, the volume of these agreements is limited to the capacities of the public finance plan	–
Relevance for the debt balance and outbalancing of liabilities and benefits: loans are on-balance liabilities and will appear in the balance sheet of the public entity. Research of the current status in the United States and Germany [2] has not shown any indication that the capital cost liabilities of DER investment have been outbalanced by the calculated benefits (energy savings etc.)	–
Risks for the obligor: The bank loan does not provide any additional services to de-risk the DER performance risks. The interest rate may be capped over time. The loan term however is fixed. Reducing the term would require extra payments	–
Risks for the creditor: The major de-risking instrument for the creditor is the creditworthiness check of the obligor, which, in the public sector, must consider the balance sheet, i.e., compare liabilities and incomes over time. The DER benefits are not considered in the balance sheet of the public entity. The performance of the DER energy savings will have an impact on the overall financial liquidity of the public body. As the public body receives revenues from taxes etc. this risk will be mitigated	+
Cost effectiveness: Transfer costs and interest rates are low as loan programs refer to the creditworthiness of the public sector and not to the specific demands of a DER	++
Matches with: soft loan programs, EPC business models (refinancing) Not combinable with: project financing	++

In the public budget of 2015 in Germany, €686 M were designated for the building refurbishment program called "CO$_2$-Gebäudesanierungsprogramm" (Building Refurbishment and Carbon Reduction Program) to support investments into energy efficiency and to boost the number of projects, e.g., with a credit product called "Energetische Stadtsanierung-Energieeffizient Sanieren," dedicated to energy efficiency for municipal buildings (effective yearly interest rate starts at 0.05%).

The financing of Energy Efficiency investment in conventional loans is often combined with preferential loans provided by national or multinational institutions at preferential conditions. The retail distribution is provided by market banks. Best practice examples are in the European context for buildings: KfW's building refurbishment program related to ambitious energy targets [3], NRW.BANK, Kredex.

Table 4.2 Assessment of eligibility for DER financing

Criterion	Eligible for DER financing?
Flexibility: The flexibility of soft loans is limited, but higher than in normal loans. The soft loan amount is fixed in the loan agreement. Often a "floor limit" (maximum loan amount) is settled using a first detailed investment cost calculation and includes, if possible a cushion for contingencies. However, the volume of these agreements is limited to the capacities of the public finance plan	+
Relevance for the debt balanceand outbalancing of liabilities and benefits: loans are on-balance liabilities and will appear in the balance sheet of the public entity	–
Risks for the obligor: The bank provides an assessment of the project to de-risk of the DER performance risks. The interest rate may be capped over time; the loan term however is fixed, reducing the term would require only small extra payments	+
Risks for the creditor: The major de-risking instrument for the creditor is the creditworthiness check of the obligor, which, in the public sector, must consider the balance sheet, i.e., compare liabilities and incomes over time. The DER benefits are not accounted, but are assessed by the soft loan bank. Risks from defaulting obligors will be mitigated by the assessment of the subsidized project and the number of different projects	+
Cost effectiveness: Transfer costs and interest rates are low as loan programs do refer to the creditworthiness of the public sector and not to the specific demands of a DER	++
Matches with: "normal" loan programs, project finance, **Not combinable with**: in some cases soft-loan programs exclude EPC and PPP	++

In the assessment of KfW preferential loan programs, the major steering instruments for energy refurbishments with more ambitious energy targets are reduced interest rates, longer maturities, and repayment bonuses (10–20%) in exchange for high efficient refurbishments, defined in source energy target values. KfW40 is a refurbishment standard that targets 40% of the minimum requirements of the current building refurbishment regulation.

The combination of a repayment bonus with decreased interest rates obviously increases the ambition; the leverage effect of public funds is usually between 4 and 10, which is higher than traditional grants. An important topic is that a preferential loan scheme allows 1:1 refinance to market banks (Basel III compliant).

A weak point in the soft loan system is that the risk adversity of commercial banks to DER projects is served with public money so commercial banks do not improve their experience with DER projects. Rather, commercial banks delegate most of the risks to the soft loan bank [4]. Since the margins are small, commercial banks do not often consider soft/dedicated loans to be a priority in their loan distribution strategy. Overall, this strategy may contribute to a first market development (Table 4.2).

4.1.3 Project Finance, Non-recourse and Recourse Finance—Refinancing of ESCO

In comparison to a loan program, the project finance (also: cash flow funding) does take into account the creditworthiness of the obligor and the transactions in which the project is financed based on its own merits. The financed project is often implemented in a project company. In the public sector, project financing is typically used to finance large scale mission related projects such as infrastructure measures, social housing, or similar large projects.

This financing format is used to refinance ESCo investments. Due to the lack of experience and performance data, "normal" DER projects that are not based on an EPC business model are not yet considered to be "revenue producing"; this may change over time with the successful dissemination of existing investor confidence programs.

The assessment of the project finance is carried out by evaluating the quality of the obligor and whether the cash flow generated will be sufficient to cover the debt service (interest plus repayment). This is done using the following KPIs:

- Debt service ratio (DSR) (net-cash flow/debt service) > 1
- Life loan coverage ratio LLC: (net present value of cash flow available for debt service/outstanding/debt in the period) > 1, at all times
- Leverage: In the public sector a leverage of 80% (loan to total capital) and 20% equity is often considered as the upper limit for loans.

Three basic types of project finance are combined:

1. Pure cash flow related finance will only take the future cash flow as security, whereas
2. Secured debt finance is additionally safeguarded by all of the project assets, including any revenue-producing contracts.
3. In recourse financing, available collateral puts the creditor in a stronger position (Table 4.3).

4.1.4 Forfeiting

Financing a forfeit means basically selling future receivables for a discounted lump sum to a bank (forfeiter), normally on the basis of bills of exchange. This financing instrument must be considered in EPC contracts between an ESCo and a public building agency to decrease the financing costs for the ESCo (Fig. 4.2). Without forfeiting, the ESCo must refinance its investment by a project finance contract with or without bank collateral; the ESCo as an industry company will be rated less reliable than a public entity and will receive a higher interest rate [5] than the public entity. With the ESCo forfeiting the future receivables (provided by the public body) to the bank, the bank will receive these receivables.

Table 4.3 Assessment of eligibility for DER financing

Criterion	Eligible for DER financing?
Flexibility: The flexibility is comparable to normal loans. Especially in refinancing ESCOs, a "floor limit" (maximum loan amount) is settled using a first detailed investment cost calculation and includes, if possible a cushion for contingencies	+
Relevance for the debt balance and outbalancing of liabilities and benefits: loans are on-balance liabilities and will appear in the balance sheet. The benefits are considered as revenues	–
Risks for the obligor: The bank provides an assessment of the project to de-risk of the DER performance risks. The interest rate is capped over time. The obligor takes the DER performance risk	+
Risks for the creditor: There are two major de-risking instruments for the creditor: (1) the creditworthiness check of the obligor, which, in the public sector, must consider the balance sheet, i.e., compare liabilities and incomes over time, and (2) the DER benefits are assessed and considered as a revenue	+
Cost effectiveness: transfer costs and interest rates are low as loan programs do refer to the creditworthiness of the public sector and not to the specific demands of a DER	++
Matches with: soft loans, ESCO refinancing **Not combinable with**: "normal" loans	++

Fig. 4.2 Forfeiting of an
EPC project

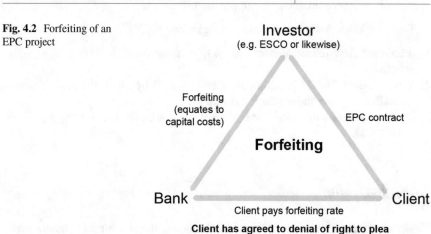

Table 4.4 Assessment of eligibility for DER financing

Criterion	Eligible for DER financing?
Flexibility: The flexibility is not necessary: forfeiting will only be used when all investment costs are available and the time schedule is fixed	+
Relevance for the debt balance and outbalancing of liabilities and benefits: Loans are on-balance liabilities and will only appear in the balance sheet of the public building owner. Aligned to an EPC project, the liability will be balanced by revenues from guaranteed energy savings achieved in the EPC project	++
Risks for the obligor (public building owner): The payment is related to a performance guarantee of an ESCo and can be considered small. In case the ESCo fails to accomplish the guaranteed savings, the public entity must fulfill the payment to the bank, but retains any rights from the EPC contract against the ESCo	++
Risks for the creditor (forfeiting bank): The obligor waives the objection; the risk is very small	++
Cost effectiveness: Transfer costs and interest rates are low; but do refer to the creditworthiness of the public sector and not to the specific demands of a DER	+
Matches with: ESCO refinancing in combination with soft loans **Not combinable with**: "normal" loans	++

- Example: A sum of €1 Million in 10 annual repayment installments, discounted at a forfeiting fee of 4% annually yields an immediate payment of €880.000 (minus approx. 0.25% provision fee, etc.)
- Passing on all accountability from the financial obligation, meaning: there is no more financial obligation from the side of the seller of the receivable (Supplier or ESCo) in case of breach of contract, non-fulfillment, etc.
- This "abstractness of the forfeiting document" will be expressed by a "waiver of objection," which means the customer waives his right to object legally against his repayment obligation because of any dispute (like non-fulfillment of conditions, late delivery, warranties, etc.).

- Forfeiture can be used when an ESCo is in an EPC arrangement with an end-user and the ESCo sells future receivables (e.g., the end-user payments) to the bank. The bank then assumes the credit risk, in return for a discounted one-time payment to finance the investments of the ESCo (Table 4.4).

4.2 Performance Based Financing Instruments

Energy Efficiency Financing Institutions Group (EEFIG) found 16 different financial instruments used for building energy efficiency in the OECD area and has emphasized the importance of the performance-related financing instruments:

- EPC
- Energy Efficiency Investment Funds
- Public ESCOs for deep renovation in public buildings
- Energy Service Agreements.

These instruments combine services with a financing instrument, in which remuneration is mainly related to life-cycle-cost related performance indicators. The business model assumes that investments are preparative measures to facilitate the performance, i.e., energy savings in EPC contracts. These instruments are dedicated to public building owners wishing to pay performance-based remuneration instead of an investment sum. The remuneration covers the capital costs of the investment and essential service costs.

4.2.1 Energy Performance Contracting

ESCo financing is a specific format of project financing with a "ring-fenced" project balance sheet that is related to the project costs and incomes. The public building owner appears in the role of contract partner of the ESCo. In most cases, the ESCo provides the pre-finance of the investment. An EPC arrangement is an integrated contract in which a contracting partner (ESCo) designs and implements ECMs with a guaranteed level of energy performance for the duration of the contract. The energy savings are used to repay the upfront investment costs, and, if agreed in EPC contract, the partition of life-cycle costs, after which the contract usually ends.

- The ESCo provides the investment and a performance guarantee; usually this guarantee relates to the energy savings and other life-cycle cost related benefits.
- After the verification of the savings in a monitoring and verification process, the public building agency is required to pay the amount of savings agreed in the EPC contract.
- In many cases, it is agreed that the beneficiary pays a fixed monthly upfront payment of 70–80% of the performance guarantee value.
- Payment for all included services delivered is based (either wholly or in part) on the monitored and verified energy savings and other defined life-cycle cost reductions.
- Typical EPC contract terms are 10–15 years, but can reach 20 (Europe) and 25 years (U.S. Federal buildings).
- Refunding of EPC investments: Depending on the available resources and on the market demand, ESCOs may finance projects out of dedicated loan programs/credit lines, or recently, out of cooperative funds and even crowd funding. Each EPC

Table 4.5 Assessment of eligibility for DER financing (EPC and Public ESCo EPC)

Criterion	Eligible for DER financing?
Flexibility: Flexibility is not necessary: EPC financing will only be used will only be used when all investment costs are available and the time schedule is fixed	+
Relevance for the debt balance and outbalancing of liabilities and benefits: In many countries, EPC is considered as a liability; it appears in the balance sheet of the public building owner. The liability will be balanced by revenues from guaranteed energy savings achieved in the EPC project	++
Risks for the obligor (public building owner): The payment is related to an ESCo's performance guarantee, which must be assessed in an M and V process. The risk to pay more than the actual savings is very small	++
Risks for the creditor (ESCO): The ESCO bears the risk that the achieved savings balance the investment and operating costs	+
Cost effectiveness: Transfer costs (to implement an EPC) and interest rates (in EPC without forfeiting) are high	–
Matches with: soft loans, forfeiting; refinancing the ESCo by project financing **Not combinable with**: "normal" loans	++

project is commonly treated as project financing that takes into account the creditworthiness of the ESCo and the NPV of the individual project. In some cases, the EPC investment is refinanced by the public entity [6]; this usually leads to a decrease in the financing costs.

- The most significant characteristic criterion between EPC and all other financing instruments mentioned here is the combination of the funding, the performance guarantee, investments, and services. In comparison to a bank loan, this would mean that the bank is taking care that their obligor is performing in a way that allows him to pay the loan amortization.
- EPC is a complex arrangement. Establishing an EPC is time-consuming and requires (external) expertise since each project must be assessed individually to estimate potential savings.
- Measurement and verification: The development of standardized and transparent procedures for M and V of savings is an important task for the promotion of the ESCo business. The key to unlocking the enormous potential for energy efficiency worldwide is securing financing. Good measurement practices and verifiability are some of the important elements in providing the confidence needed to secure funding for projects. Securing financing requires confidence that energy efficiency investments will result in a savings stream sufficient to make debt payments. M and V practices allow project performance risks to be understood, managed, and allocated among the parties. Currently IPMVP [7] is a standard methodology in use at least for the markets in the United States, Canada, and UK (Table 4.5).

4.2.2 Public ESCOs EPC for Deep Renovation in Public Buildings

A best practice example of this instrument is a public ESCo in France called OSER. It is a local public company (LPC) based on a private structure with the purpose of providing energy efficiency services dedicated to implement deep retrofit projects in public buildings in the region of Rhône-Alpes. The owners are 10 municipalities of the Rhône-Alpes region, the Intercommunal Union for Energy Issues in the Department of Loire (Le Syndicat Intercommunal d'Energie du département de la Loire), and the region of Rhône-Alpes. A benefit of this model is that the complex tendering processes of "normal" ESCOs can be avoided by using the public ESCo.

The public ESCo uses a business model of a shared savings structure with the municipalities. Deep retrofit projects use 10% down equity of the municipalities. The LPC will optimize the overall costs of the project including the financial costs. The partial repayment of the investment is by a fixed rent paid from the municipality directly to OSER. For a performance guarantee, the LPC agrees with a commercial ESCo or subcontractor to assign special services and operational activities (design, implementation and maintenance) in an EPC.

The public ESCo generally requires a short-term loan to start the project. This can be provided by public banking institutions (EEEF, Caisse des Dépôts et Consignations, etc.). In this business model, in terms of refinancing the projects, the use of another (long term) loan is enabled with a new negotiation, which will carry out the assessment and will look for the proper loans and other financing sources. However, the disadvantage of this instrument is the impact on public debt and limited experience of the public ESCo.

4.2.3 On-Bill Repayment (OBR)

In an OBR, the customer and contractor identify and realize viable EE-potentials in the building. After the verification of the project performance (installation of EE measures), the customer's loan is repaid by monthly surcharges added to the utility bill. Since energy costs decrease due to the investment in EE, the utility costs remain on the same level or are even reduced. The risk of default of the building owner is not completely eliminated, but the debt collection is facilitated as it is connected to the energy consumption of a building. As soon as energy is consumed, the OBR becomes a part of the account. Also the debt service is connected with the building, not with the building owner. In case of transfer of a building, the debt service is handed over to the new proprietor (Fig. 4.3).

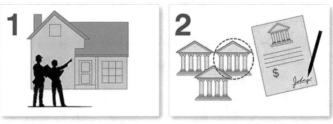

Fig. 4.3   On-bill repayment

4.2.4   Conclusions

The assessment of a number of selected financing instruments with regard to specific requirements of DER financing shows that:

- In comparison to other financing instruments, the bank loan provides high risks for the obligor as the flexibility (total amount and time schedule) is low. The performance risk of the DER investment must be taken by the obligor and is not considered in the creditor's due diligence process. The creditor can rely on the good creditworthiness of the public entity. If combined with EPC business models (where the ESCo manages the risks mentioned before), the loan can be a highly attractive funding source.
- Soft loans may reduce some of the obligor´s risk as they provide more flexibility in terms of redemption, lower interest rate-risks, and, in most cases, a due diligence of the proposed DER project.
- Project finance combines the creditworthiness of the obligor with the quality of the DER project cash flow. Thus, the project finance is not much in use for appropriately funded DER projects in the public sector. Project finance (PF) is one of the instruments to refinance EPC projects with the ESCo as an obligor.
- The forfeiture mitigates the risks of EPC financing models between the ESCo and the public entity: by waiving the rights of objection, the ESCo can drastically

Table 4.6 Criteria for DER financing instruments

Criterion	Loan Financing (LF)	Soft Loans (SL)	PF	Forfeiting	EPC/Public EPC
Flexibility	−	+	+	+	+
Relevance for the debt balance	−	−	−	++	++
Risks for the public building owner (obligor)	−	++	+	++	+
Risks for the creditor	+	++	+	++	+
Cost effectiveness	++	++	++	+	−
Combinations with other financing instruments (FI)	++ with EPC, SL	++ with loan, PF	++ with EPC, SL	++ with EPC, SL	++ with Loans, PF

reduce the interest rates of the EPC project, which in turn increases the cost-effectiveness of DER EPCs (high investments) (Table 4.6).

References

1. Report from the European Commission to the European Parliament and Council on alternative tools to external credit ratings….Brussels, October 2016 COM (2016) 664 Final: while public bonds in the international context are sued by a by a sovereign national government, the municipal bonds are issued by local governments to finance public projects such as roads, schools, airports and seaports, and infrastructure-related repairs. The term municipal bond is commonly used in the United States, which has the largest market of such trade-able securities in the world. As of 2011, the municipal bond market was valued at was at $3.7 trillion; UK: the UK municipal bonds agency was set up in 2016 as the first for the sector in the UK. It issues bonds to finance local authority projects at a lower cost than the Debt Management Office. This lowers council's finance costs, which means more can be invested into local economies, infrastructure and housing projects. The Debt Management Office runs a Public Work Loan Board which is managing to provide loans and to collect the repayments
2. EDLIG, Workpackage B Report (German): Assessment of 16 DER Projects from the Perspective of Finance, Karlsruhe (2014)
3. https://www.kfw.de/KfW-Group/Newsroom/Aktuelles/Pressemitteilungen/Pressemitteilungen-Details_10591.html
4. EEFIG Final Report, Brussels (2015)

5. IRENA Executive Strategy Workshop: Practical Policies for Financing Renewable Energy Action Plan Investments, Brussels (2015). Currently a 20 years loan interest rate will be between 1.2–2% for a public body and 3.5–4.6% for an ESCo (without forfeiting)
6. Einsparcontracting mit Eigenmitteln, Lohse, Karlsruhe (German, 2008)
7. www.IPMVP.org

Chapter 5
DER Business Models

5.1 Business Models—Definitions and Introduction

The business model is describing a product's value proposition, infrastructure, customers, and finances. It assists firms in aligning their activities by illustrating potential trade-offs. In this case, the product is the DER in the public building sector. The target of the business model assessment is to understand the services, mechanisms, and incentives of each of the party involved.

Currently, the implementation of the DER is, in the majority of cases, carried out in an "owner-directed" business model. The distribution of services and responsibilities, combined with false incentives in the remuneration system and the requirement for public entities to account for the liabilities (capital costs), shows deficits and does not contribute to an increase in the number and pace of DER projects in the public sector.

The link between the business model, limited appropriate funding in the public sector, and the increasing need for private money is the **bankability of the benefits**. A benefit can be considered as a value for out-balancing the capital costs (of a loan, soft loan, PF, or EPC) if the benefit is trust- or credit-worthy. This is given as soon as the benefit is measurable, as soon as it can be verified, and as soon as it is generally accepted by the building owner.

As soon as this is given, the value of the benefit can be linked to the liability (capital costs). In a best case scenario, the bankable benefits outbalance the liabilities. To date, only EPC can provide bankable benefits; however EPC has not been in use in DER projects.

Two recently developed approaches may contribute to improve the owner-directed business model toward bankable DER benefits: the ICP Europe program provides transparent project performance protocols for each stage of a DER project. The SMESCo program, however, works with a customized due diligence process that helps to carry out a standardized due diligence process and to provide loan guarantees for SMESCo investments.

R. Lohse and A. Zhivov, *Deep Energy Retrofit Guide for Public Buildings*,
SpringerBriefs in Applied Sciences and Technology,
https://doi.org/10.1007/978-3-030-14922-2_5

For the EPC-business models, four replicable approaches to DER EPC business models are available for the public sector:

- In the United States, GSA implements EPC business models that combine public and private sector money to carry out DER. First results show EUI reductions of more than 40%.
- In Belgium Factor 4 integrated a simple mechanism to include the added asset value, the improved building indoor climate and the high value maintenance program into the remuneration structure of an advanced DER EPC.
- RenESCo is executing innovative EPC for DER in the housing sector in Latvia by integrating energy and maintenance cost reductions into the EPC financing scheme. The refinancing of RenESCo is provided by the collected and forfeited savings and soft loan programs of the EU.
- In Germany KEA has implemented a DER EPC by including energy savings, energy substitution by renewables, and maintenance cost savings into the financing scheme. The building pool was designed by one DER and four "normal" HVAC assets to mitigate the performance risks.

The following section briefly introduces the major business models. Further information such as budgetary details and contract stipulation is available in advanced DER business model case studies described in the appendix.

5.1.1 *Appropriated Funding and Execution Model*

The first and currently most popular model in the public sector is based on the **use of appropriated funds available to public/government building owners** within the setup of national, municipal, or regional budget planning. Government agencies or municipal administrations are responsible for the budget planning and for the execution of the investments in their building stock. The budget may include public equity (tax payments etc.) and dedicated bank loans. In most European countries, however, bank loans are limited by a public debt ceiling that is related to the available equity of the public body.

In the budget planning, the building refurbishment competes with other tasks that a public entity must fulfill. Building refurbishment is usually not one of the first priority topics on the national, regional, and municipal levels. Thus, the appropriated model often has limited appropriated funds to renovate existing buildings, whether to repair aging infrastructure, update building interiors, plan for disaster preparedness and resilience, or perform energy upgrades. Agencies typically have some funding available for specific building improvements under programs like (in the United States) the Department of Defense´s sustainment, restoration, and modernization (SRM) program.

From time to time, the public sector creates specific employment and market stimulation programs that make a larger extra budget available for governmental, regional/state, and municipal building owners. This has been the case in the follow up to the financial crisis of 2008. In the United States, Federal buildings benefited from appropriations delivered under the American Recovery and Reinvestment Act of 2009 (ARRA), which awarded $5.5 billion to the GSA and $7.4 billion to the Department of Defense for the construction and renovation of buildings for energy efficiency improvements and other modernization efforts. In Germany 2008 and 2009 two initiatives "Konjunkturpaket I/II" were established to stimulate the market by public investment spending of some €5.9 billion in public sector building refurbishment.

In this model, building owners take responsibility for the project design, and for the management and financing of an energy efficiency retrofit to their property. They take full responsibility, and assume liability, for the quality of the project and for the economic return on their investments. The building owner controls contracting, retrofit component selection (and hence the retrofit project price), and project management. The owner is fully liable for the retrofit's subsequent economic performance (i.e., volume of energy required to deliver post-retrofit living conditions), and for the financing (which is possibly secured), but not directly for the retrofit components or for its overall energy performance. By assuming the risk for all the retrofit components, the building owner is well placed to benefit from any economic outperformance (i.e., when energy prices go up faster than planned) and can clearly benefit directly from a higher grade Energy Performance Certificate, and from improved acoustics and livability. The current contract and remuneration model does not provide incentives to the planners, architects, and craftsmen to provide high energy and cost efficient project structures, technologies, or methods of implementation. In some countries, as in Germany, the remuneration system of architects is designed in a way in which the remuneration increases with the complexity and the investment costs; the relation between payment and investment costs creates a counter-productive effect on the cost-effectiveness of EEMs. Beyond that, the appropriate directed business model has several serious shortcomings:

- The feedback model is "open," i.e., there is no feedback based on operational experience. This influences the quality of planning, construction, and operation.
- Decision making is fixed to one key criterion, initial investment, which does not account for LCCs.
- Neither planners nor architects are required to provide follow up or respond to questions related to energy performance or the investment costs.

Currently, the experience derived from the performance of DER projects is not collected, evaluated, or distilled into lessons learned. In other commercial or industrial settings, the business process would follow well defined steps that would include a "feedback loop." The experience of DER projects would be documented and evaluated and its performance measured and analyzed. This analysis would be used to produce lessons learned, which would be implemented in subsequent projects. Over time, this evolutionary process would improve the business model. The building sector would benefit from adopting these steps. All this is barely taking place in the public sector.

The uncertainty, which results from a simple lack of information, has two major impacts: (1) building owners are reluctant to believe that DER may contribute significantly to the performance of their buildings (which restricts demand), and (2) private money does not find its way into these projects. Overall, it seems unlikely that the ambitious EU 2020 targets will be practically achieved in the EU as a whole, or on the level of any single nation, until this central problem is resolved:

> How can the DER market measure and verify the performance of investments in DER while applying the owner-directed business model?

5.1.2 Fixed Payment Model and Utility Fixed Repayment Model

The second model (so far primarily used by commercial building owners) is a **fixed repayment model**, in which the upfront capital cost of an energy efficiency retrofit is organized, subsidized, and at times fully provided by either utility or through a PACE (Property Assessment Clean Energy) program financing mechanism established by a city, county, or Port Authority in the United States. These investments are repaid through monthly, fixed, non-performance-related surcharges. The "**Utility Fixed Repayment**" version of this model requires a supportive policy framework to function; the types of legislative changes that regulators have may include: requirements for electric and gas utilities to improve their customers' energy efficiency by a certain amount each year; the application of white certificate programs or the decoupling utility profits from the quantity of electricity sold; and requirements that utilities invest first in the lowest cost sources of energy.

Although the remuneration of the utility is not related to the actual performance of an implemented project, the **Utility Fixed Repayment Model** has several immediate advantages over the **Appropriated Funding Model**:

1. Utility cost of finance, access to funds and available leverage should be considerably better than that achieved by owners under **Appropriated Funding Model**.
2. Friction costs are reduced from the economies of scale created by a utility executing many hundreds or thousands of its individual client retrofits.
3. Customer "ease of execution" is enhanced as execution is streamlined and there is less work for the building's owner than in Owner Financed Model.
4. Government can use its relationship with the Utility sector to align interests and push national energy efficiency targets down to the corporate level through the imposition of Standards and Market-Based programs like CERT in the UK or the white certificate scheme in Italy.

5.1.3 On-Bill Repayment Program

One of modifications of the Fixed Repayment program (OBR) is offered by Environmental Defense Fund (EDF) in several states in the United States. It can work for single-family, multi-family and commercial buildings. It can also work for both tenant- and owner-occupied properties. OBR can accommodate a variety of energy-saving opportunities including equipment purchases, equipment leases, Energy Service Agreements, and Power Purchase Agreements. While on-bill financing refers to programs that use ratepayer, utility shareholder, or public funds, OBR programs leverage private, third party capital for financing. Banks, credit unions, or financial institutions provide the loan capital and loan payments displayed on utility bills. This approach allows third party institutions to take care of administrative functions, while utilities need only process payments. OBR obligations can use several different financing vehicles, including loans, leases, and power purchase agreements (or PPAs, which serve as agreements to buy and sell energy savings over time).

Property Assessed Clean Energy (PACE) is a modification of fixed repayment model financing mechanism that enables low-cost, long term funding for energy efficiency, renewable energy, and water conservation projects www.pacenation.us/commercial-pace. PACE financing is repaid as an assessment on the property's regular tax bill, and is processed the same way as other local public benefit assessments (sidewalks, sewers) have been for decades. Depending on local legislation, PACE can be used for commercial, nonprofit (public), and residential properties. PACE can cover 100% of a project's hard and soft costs with financing terms up to 20 years. It can be combined with utility, local, and federal incentive programs. Energy projects are permanently affixed to a property's tax bill, stay with the building upon sale, and are easy to share with tenants.

Figure 5.1 shows the major functionalities of the PACE financing model:

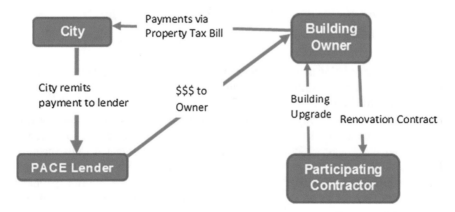

Fig. 5.1 PACE financing scheme. *Source* Pacenation.com

1. The city, county, or Port Authority creates Financing District.
2. The property owner voluntarily applies for financing (which is typically combined with Utility or other incentive programs).
3. Proceeds from financing are provided to property owner to pay for project.
4. The property owner installs projects and repays the loan through property tax bills (up to 20 years).

5.2 Energy (Saving) Performance Contracting (Private Funding) Model EPC/Energy Savings Performance Contract (ESPC)

From the financing perspective, the EPC is the only currently established model in which an energy efficiency retrofit provider designs a retrofit and finances it, and is repaid only through the cost savings resulting from energy and other life-cycle cost savings, therefore assuming the responsibility for the economic success and quality of the retrofit. Performance contracting, typically delivered in the form of EPCs (ESPCs) or UESCs (utility energy savings contracts), allows public building authorities to deliver energy savings independent from the normal public investment budget as it turns operative costs (OPEX) into capital investment costs (CAPEX) without special appropriations. Project costs are financed by an energy service company. In some cases, financing is provided by the public building owner and paid back over time based on the guaranteed, measured, and verified energy savings of the project. So far, the normal ESPC is not in use for DER projects. The following section further explores the structure of ESPCs (Fig. 5.2).

5.2.1 Blended Funding (Public and Private Combined Funding)

This model is a variation that can be described as an upfront payment by the building owner integrated in an ESPC business model that can improve the economics by reducing the total cost to be financed. The upfront payment is used to enable the integration of cost-ineffective measures into the scope of an ESPC without making the ESPC for both parties unattractive. In the United States, there is a long history of agencies using appropriated funds, including energy-designated Department of Defense funds, as one-time payments in ESPC projects. There is often a strong argument for applying funds designated for non-energy projects as a one-time payment for an ESPC project to drive greater value, but the legal limitations of combined funding models must be considered.

To maximize the value of DERs, agencies must both understand the opportunity of pursuing a DER with combined funding sources, and be prepared to act when the

CYCLE OF COST SAVINGS AND PAYMENTS

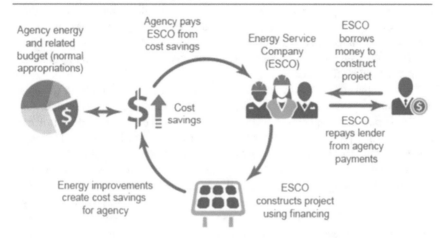

Fig. 5.2 A typical U.S. Federal ESPC project structure. *Source* DOE, FEMP 2016

timing is right. Developing an energy master plan is the key first step to understanding the opportunities that a site may offer. It is crucial to have an independent third party to lay out the site's energy master plan so the agency can have a neutral opinion that can inform requests for appropriated funding and potential ESPC projects over time. This energy master plan should be closely coordinated with an energy capital investment plan so that an agency can be prepared to execute and fund energy-related projects appropriately as funding becomes available. Additionally, the energy master plan should remain flexible to pursue combined funding projects as energy-related funds become available.

5.2.2   U.S. Precedent for Combined Public and Private Funding

There is currently precedent for combining ESPCs with appropriated funding in situations where that funding has been specifically designated for "related" projects, where the appropriated funds are intended for energy-related projects. U.S. Department of Energy has guidelines regarding one-time payments and one-time savings or cost avoidance in ESPCs that was accepted by the Federal ESPC Steering Committee on December 5, 2006. These guidelines explain how appropriated funds can be applied to an ESPC. The guidelines apply to projects that are solicited and awarded as an ESPC. The law, 42 USC 8287, has a provision that allows for some appropriated funding to be applied to an ESPC. This enabling legislation provides that ESPCs are for the purpose of "achieving energy savings and benefits ancillary to that purpose."

It also states that payments to an ESCO "may be paid from funds appropriated or otherwise made available to the agency for fiscal year 1986 or any fiscal year thereafter for the payment of energy, water, or wastewater treatment expenses (and related operation and maintenance expenses)." It is imperative that the appropriated funds that are going to be applied to an ESPC be directly related to the energy measures being executed by the ESCO.

For example, if an agency had funding available that was intended to replace existing single-pane windows with slightly more efficient double pane windows, an ESCO, as part of an upcoming ESPC, could finance the incremental cost of more advanced triple pane windows that will further reduce building loads. The appropriated funding for the original window replacement could be applied to the ESPC as a one-time payment that would drive greater value from the window replacement through added energy savings and overall project cost effectiveness. This can result in an increased scope for the ESPC project (or in an opportunity for the agency to buy down the ESPC contract term). If this project is timed with the trigger of central HVAC system replacement, the reduced heating and cooling loads from the triple pane windows could allow a less expensive, lower capacity HVAC system replacement. These synergistic approaches are what enable the 50% savings that can be achieved in deep retrofits [1].

There are challenges in combining energy and non-energy projects. While a combined funding approach can deliver deeper savings on limited budgets, several barriers prevent broad implementation of this model for the U.S. Federal Government agencies. These limitations do not apply to other cases including state and city government projects. In federal contracts, ESPCs can only be paid from the savings that are generated from work that is executed as part of the ESPC. When an installation receives appropriated funding for an SRM project, that project is supposed to be solicited based on the rules in the Federal Acquisition Regulations (FAR). This process can, but does not currently consider the potential to combine an ESPC effort with the SRM "funding" that could be used for "related" (energy-related) projects. If there is no relationship between the ESPC projects and the "funded" project, the FAR would prevail and the non-energy-related scope would be solicited separately from the ESPC efforts (Fig. 5.3).

Soliciting non-energy-related scope separately from the ESPC efforts would significantly complicate the project's efforts. From a logistical standpoint, having two or more contractors onsite implementing closely intertwined scopes adds significant complexity to project implementation. Client teams would need to coordinate two contractors with different contracts, schedules, sub-contractors, and scopes to work together in the same space, at the same time, without adversely impacting the project as a whole.

Figure 5.3 shows the schematic of the combined funding model in which the General Contractor (GC) constructs the entire project, but the energy-related portion is implemented under a subcontract with the ESCO. The GC has two managers (government customer and ESCO), but the government customer ultimately is in charge of entire project.

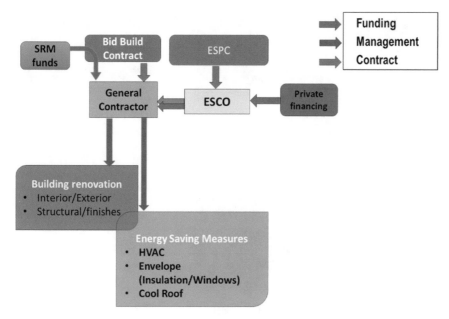

Fig. 5.3 Schematic of the combined funding model

The major legal limitation is not necessarily identifying what scope can be performed by the ESCO under an ESPC. The legal limitation relates to whether or not an agency can advertise a "funded" project as an ESPC, since by law, ESPCs are third party financed arrangements. Generally, an ESCO may only perform energy- or water-related conservation measures and related ancillary construction (such as concrete pads under and enclosures around equipment) and operations and maintenance work. If a "funded project" is solicited to an ESCO group, it is likely that the contract community that normally bids those types of projects would protest that the work is not ESPC work. However, current rules would allow an ESCO that is performing related work to use funding as a one-time payment (for agency cost avoidance) if the funding becomes available to use during the right stage of ESPC development. However, the challenge of timing remains significant. Early communication and awareness at an agency or installation regarding projects that could build upon each other to achieve savings is key, but there is always an underlying risk that planned funding will not be made available.

Potential contractor arrangements. There are many challenges associated with having separate contractors working on the respective energy and non-energy project scopes. This collaboration could take many forms. In one instance, an ESCO could serve as a subcontractor to a prime contractor delivering non-energy services as part of the SRM project. In this scenario, the agency would not have any privacy with the subcontractor, so they would have to work through the prime contractor. Also, the agency's relationship with the prime contractor would likely be awarded as a

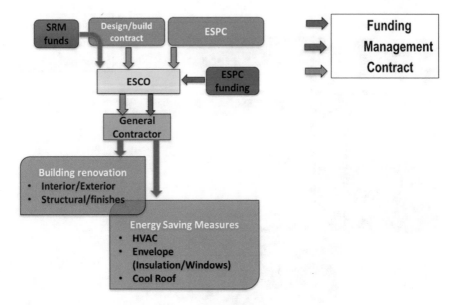

Fig. 5.4 Schematic of the combined funding model

construction contract, an operations and maintenance contract, or a service contract, which could include some construction effort. Those types of contracts would be subject to the FAR, and can generally be in place for only 5 years. This would prevent the agency and the ESCO from benefitting from the partnership of up to a 25-year contract term, which is necessary to deliver substantial energy savings as part of a DER. There are no regulations in place that can bridge the gap of the agency's ability to work with the subcontractor.

There are also challenges if the ESCO is the prime contractor and the agency is trying to incorporate the SRM project or project funding in with the ESCO work (Fig. 5.4).

Figure 5.4 shows the schematic of the combined funding model in which an ESCO is awarded a design/build contract for non-energy-related building renovation, and an ESPC for energy-related measures. The ESCO hires a GC, but provides a single point of contact for the government customer.

There has been ongoing discussion to evaluate methods that could be used where an ESCO is in place and has the potential to add value to SRM work. One potential option could be for the ESCO to provide equipment to a prime contractor as government furnished equipment. There are several challenges with how this could transpire, since the SRM contract assumes that the funding covers the entire project (including energy and non-energy scope). The ESCO and an SRM contractor would have to work out the specific arrangements that would allow for this to happen—ensuring that neither contractor performs work outside of the scope of their respective contracts. There could also be challenges during the operation phase of the ESPC if the ESCO alleges that the provided equipment was damaged or not properly installed by the SRM contractor, and that this is the reason that savings are not being realized. So, there are many challenges when separate contractors are hired to perform related energy and non-energy work on an SRM or similar project.

In summary, there are legal issues with how a contract can be structured to comply with 42 USC 8287 and to still not violate the FAR if appropriated funds are anticipated to be available at the time of contract award. There are privacy of contract issues if the ESCO is a subcontractor to a prime on an SRM project, which would inhibit the agency's ability to accept a comprehensive ESPC project from the prime. There are also issues related to an ESCO performing work that is not energy work. Some limited non-energy work could be allowed, but substantial non-energy-related work performed by the ESCO or a subcontractor to the ESCO would not be allowed. So, it is critical that, if there is a potential project that could achieve greater savings using the DER concept, the team evaluating that project know and understand the procurement rules, and clearly delineate the energy and non-energy scopes to bring the greatest value to the ESPC project.

5.3 Advanced DER EPC Business Model

The link between the business model, limited appropriate funding in the public sector, and the increasing need for private money is the bankability of the benefits. A benefit can be considered as a value for out-balancing the capital costs (of a loan, soft loan, PF, or EPC) if the benefit is trust- or credit-worthy. This is given as soon as the benefit is measurable, can be verified, and is generally accepted by the building owner.

As soon as this is given, the value of the benefit can be linked to the liability (capital costs). In the best case, the bankable benefits do outbalance the liabilities.

Currently, only a few business models can provide bankable benefits. The prerequisite of accountability is a guarantee or other strong proof of evidence for the benefits. A performance guarantee is stipulated in ESPCs or EPCs.

This contrasts with the owner-directed business model, which does not provide bankable guarantees (nor do leasing, PPP, in-house contracting etc.), where no enforceable responsibility for the energy or maintenance performance is in place. The following two sections present practical approaches to advance benefits from DER projects implemented in an "owner-directed" business models into a bankable approach. For the EPC business models, four approaches toward DER EPC business models are presented: (1) U.S. GSA, (2) Belgium Factor 4, (3) Latvia RenESCo, and (4) Germany KEA.

How to Improve the EPC Toward DER EPC

EPCs are proven tools that guarantee energy and maintenance [2] cost savings and that provide some essential security in comparison with other business models. These tools can provide:

- Strong contract-based stimulation for both contract parties to achieve high cost effectiveness by providing a better savings/investment ratio.
- Guaranteed energy and maintenance cost savings between 25 and 40% in both the United States and the EU.
- Bankable energy and maintenance cost savings, which create reliable revenue streams to fund deep retrofit projects.
- Cost structure and decision-making criteria aligned with LCCs.
- Energy Service Company's (ESCo's) design and experience based knowledge on different ECM bundles (e.g., HVAC/biomass/CHP, etc.) that give satisfactory performance results.

EPC is currently not the chosen vehicle for DER projects. The key strength of major ESCOs is still in the building automation. In some countries, the scope of EPC has already been extended to include renewable energy and some infrastructural measures.

Some additional elements are needed to provide the necessary framework to prepare ESCOs to enter into DER EPC pilot projects to advance existing EPC business models for DER projects:

- Creation of financing schemes that can **integrate revenue streams** deployed by energy, maintenance and other life-cycle cost savings.
- Assessment of national framework conditions for public building owners to account for increased **residual building values** provided by a DER project.
- Quantification and integration of non-energy or non-cost related benefits into the cash flow analysis to monetarily quantify all DER benefits in addition to the energy-related benefits.
- Creation of a viable database for DER projects to collect evaluated data for QA.

Market and Framework Conditions

It can be a challenge to direct ESCOs to take constructive (profit-making) measures that fall outside their core business activities; such actions can require marketing support and may even affect the conditions of the ESCO framework. In Germany [3] an initiative, and associated working group DER EPC, has been established to discuss and prepare necessary steps:

1. The market development for DER EPCs must be carried out jointly by building owners, ESCOs, facilitators, financiers, and technical experts. These stakeholders must be prepared to enter into this endeavor as a constructive, cooperative activity.
2. Participants must carefully assess the DER process to mitigate risks when preparing national standard contracts, project development and implementation structures, award criteria, tendering processes, etc. It is important to bear in mind that specific investment costs of a DER measure bundle will be two or three times higher than the costs of a "normal" HVAC measure bundle, which will consequently lead to extended payback and EPC contract periods.
3. Over a longer contract period, the allocation of risks and responsibility, which can be found in most European standard EPC contracts, will lead to an escalation of risks:

    a. Maintenance and replacement cost for the building automation, heating pumps, motors, CHP units, heat pumps will increase.

    b. Refinancing costs will increase or will at least be difficult to predict.

    c. Over long contract periods, monitoring and verification will become even more complex as changes of utilization and floor space become more likely.

4. A major target of this effort must be to mitigate risks resulting from long contract periods which is a major concern of ESCOs, building owners, and financiers. The following key measures have been implemented successfully:

    a. Integrate additional life-cycle cost benefits to reduce the payback and contract period. In this case study, avoided maintenance and replacement costs for the existing equipment and constructions will be refurbished by the ESCo. This approach added another 25-30% to the energy savings.

    b. Optimize refinancing for ESCOs by providing long-term refinancing sources with stable interest rates such as energy cooperatives or green funds. If a critical investment cost level can be reached, DER EPCs would be eligible for pension funds.

    c. Optimize investment cost-benefit ratio by establishing a building pool with short-, mid- and long-term payback periods.

5. Award criteria: The modification of the normal EPC award criteria to giving a monetary value to sustainable technical concepts can improve the cost-effectiveness in certain areas. In this case, the ESCo provided a bid that specified an overall U-value 15% better than that of the minimum requirements. After the ESCo bid was evaluated, the award criteria allowed an increase in the number of credit points since this additional rating equates to 2% of the savings criterion. This, when combined with the additional energy savings resulting from the better U-value, covered the additional capital costs of the related additional investments. From the perspective of the building owner, the added value of a better U-value or a more resilient mechanical system lowers maintenance and replacement costs over the building's lifetime.

The necessary adjustments resulting from this analysis have been gathered into a set of de-risking measures (DRMs) in a revised version of the German EPC contract and project structure templates. This is considered to be a first step. Follow-ups will be necessary after the conclusion of the first and second implementation phases, and after the results of the first year's performance and of the first M&V process have been gathered. In the meantime, the project team will work on the refinancing sources for ESCOs, which is seen to be a crucial factor for the future success of DER EPCs.

5.4 QA Tools for Investors

In addition to the measures in the previous steps, the QA process can be further improved by using databases to crosscheck the DER project data and by considering performance protocols like ICP Europe.

5.4.1 Data Sources: Investment and Energy Performance

Information is needed at the project level on the investment cost, the energy and LCC performance of single measures, and measure bundles combined with services. The following online tools, which can provide insights for prospective energy efficiency projects and investors from different sectors, can serve a role model for the public sector:

- The EEFIG project II [4] is currently designing a database for EEMs in buildings with information on the investment costs and performance of single ECMs and small ECM bundles.
- The project ENTRANZE [5] database, which is based on hourly modeled net zero energy (NZE) approaches for three different building types in six different climate zones in Europe. The investment costs refer to national investment cost values. Results of the calculations show the optimum investment costs for site energy performance of assessed building types.
- The German Database for Investment Cost Calculation (design-and-cost database, PLAKODA) is based on an extensive cost-evaluation of energy-saving measures carried out within the context of the federal funding programmes "Energieeinsparprogramm" (energy-saving-programme, EEP) and "Konjunkturpaket II" (economic-growth programme, KPII). This database has been established to facilitate investment cost forecasts, especially in the public sector. The specific cost values and functions derived in the present study allow such cost forecasts for typical energy-saving measures on thermal building envelopes as:
  - thermal insulation of outer walls, floors, roofs/attics
  - installation of new windows
  - measures on building systems, e.g., installation of heat generation systems with wood boilers or condensing gas boilers,
  - thermal solar collectors and pipeline systems.

- Energy Intensive Curve (£300 mm of mainly UK EE investments); source: The Crowd (2015). Energy Investment Curve. [Website]. Retrieved from: http://thecurve.thecrowd.me/.
- Green Button (U.S. database with energy use data for 60 million customers used for benchmarking in commercial and residential buildings sectors) retrieved from: http://www.greenbuttondata.org/.

- IIP's Industrial Efficiency Technology Database (Global research and benchmarks for cement, iron, steel and pulp & paper sectors plus electric motor driven systems) Institute for Industrial Productivity (2015). Industrial Efficiency Policy Database. [Website]. Retrieved: http://iepd.iipnetwork.org/.
- Investor Confidence Project (containing financial performance data for 12,000 U.S. home energy efficiency loans) Investor Confidence Project (2014). Enabling Markets for Energy Efficiency Investment [Website]. Retrieved from: http://www.eeperformance.org/.
- U.S. Department of Energy supported Industrial Assessment Centers Database (containing 16,700 assessments and over 120,000 recommendations).
- Pilot open-source, data and benchmarking EEII platform containing macroeconomic energy performance data, survey responses from audits, management systems and performance certificates for industrial energy efficiency, which has the potential to identify and to address challenges on a continuous basis and allowing for cross-country comparisons.
- Investment performance: Especially for large building stock management funds, it is important to provide performance indexes. In Australia, the performance analyst Investment Property Databank [6] (IPD) compiles an Australian Green Property Index, which tracks the investment performance of A\$53bn (£34bn) of commercial office buildings that have been awarded a Green Star environmental performance rating. Until June 2012, this benchmarking tool reported that Green Star rated offices delivered an annualized return of 10.6%, which surpassed the all-office sector's return of 10.5%. In the UK, a lack of investment-related environmental information means that it has never been clear whether sustainable properties deliver a better return. Now, with the recent launch of IPD's Eco-Portfolio Analysis Service (EcoPAS), experts are predicting that the same differentials will start to become apparent in UK and France.

5.4.2   The Investor Confidence Project (ICP) Approach

The Investor Confidence Project (ICP Europe) has defined a clear roadmap from retrofit opportunity to reliable Investor Ready Energy Efficiency™. With a suite of energy performance protocols, ICP reduces transaction costs for building owners, ESCOs, and other institutional investors who need to assess the technical and financial viability of DER projects. The ICP's energy performance protocols assemble national standards and practices into a consistent and transparent process that leads to comparable results. The goal of this project is to develop standardized protocols that describe how to execute certain crucial processes:

- Baseline definition
- Savings projection
- Design, Construction, and Commissioning
- Operation, Maintenance, and Monitoring
- M&V.

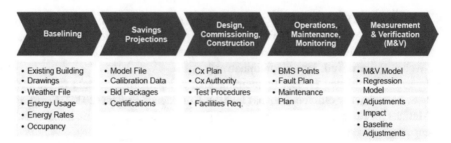

Fig. 5.5 ICP DER project assessment

A certain level of QA is implemented through standardization. All standards are documented in the Energy Performance Protocols (Fig. 5.5).

Thai a not meant to imply that a positive assessment will guarantee that a labeled project will have a confidence level of 100%. However, it does serve as a basis for an investor's risk assessment and will therefore help to reduce risk add-ons (i.e., increased interest rates). DER projects that use the energy performance protocol are rewarded with a certificate.

The ICP concept has been implemented with in ST B research work on a DER project in Mannheim. In this context, the first three protocols ("Baseline," "Saving Projections," and "Design, Construction and Commissioning") have been used for the first time in Germany.

5.4.3 SMESCo Business Model

In the context of the German National Initiative for Energy Efficiency, a work group representing a collaboration of the German DOE, German guarantee banks, and KEA designed a business model that address the needs of small- and medium-sized ESCOs (SMESCos) for energy efficiency investments in small- and medium-sized assets. The business model (Fig. 5.6) is built on a PF instrument and is combined with a loan guarantee provided by a public bank. In 2 years, the business model will be able to implement the first DER projects.

Scope

In the three-stage qualification and execution program, designers, architects, SME contractors, and tradecraft companies are qualified to provide EPC projects to their customers.

Financing Instrument

1. The SMESCo will contact their business partner bank to finance the energy efficiency investment either by project financing or by a traditional loan, combined

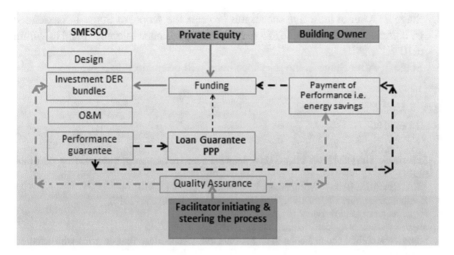

Fig. 5.6 SMESCo business model (Lohse 2016)

with a soft loan program and subsidies. The SMESCo will provide a cash flow plan for the project and credit rating data.

2. In most cases, the business partner bank will not be able to carry out the technical and commercial due diligence process of an energy efficiency project. In that case, the business partner bank will contact the Regional Guarantee bank to conduct the due diligence and to provide a partial loan guarantee.

3. The due diligence will be conducted via collaboration between the guarantee bank and experts with an established record of experience EPC in evaluation by means of a customized evaluation tool. At the end of the process, the project will be rated and the business partner bank will provide a partial loan guarantee to secure the project financing.

4. The loan guarantee extends the SMESCo´s equity rate since it is not considered a liability.

Process

1. Stage 1: By signing a normal design or construction contract, the contractors/designers obligate themselves to design and implement an energy efficiency measure that provides a stated level of control of energy consumption, and a recommissioning of installed equipment. The goal here is to minimize the performance gap between the predicted energy savings and the actual performed energy savings. Controls will be provided on the basis of a remote control system that provides access to the utility meter. To mitigate risks and to maintain a high quality service, the experts will provide assistance in managing the controls and the recommission. The performance results and the quality of the SMESCo´s services will be evaluated and recorded in a public database.

2. Stage 2: After at least five successful projects the scope, of Stage 1 is extended by a performance guarantee to a performance guarantee tolerance band of within ±10%.
3. Stage 3: After Stage 2, the tolerance band will be reduced to 5%.

References

1. Experience with DER ESPCs in the USA, John Shonder, IEA Annex 61 Investors Day, Karlsruhe (2014)
2. KEA, Karlsruhe Baden- Württemberg EPC Model Contract (2010)
3. Kompetenzzentrum Contracting, DER Working group, Karlsruhe (2015)
4. http://deep.eefig.eu/overview
5. http://www.entranze-scenario.enerdata.eu/site/
6. http://www.building.co.uk/ecobuild-and-sustainability/ecobuild/getting-to-grips-with-micro-economics/5045225.article

Chapter 6
Conclusions and Outlook

The overwhelming majority of building renovations are "shallow refurbishments" that focus on single measures, e.g., HVAC replacement, and thereby tend to miss critical opportunities to make much needed envelope improvements, such as façade upgrade, or roof or window replacement. In practice, HVAC measures are rarely combined with refurbishment of the building envelope, and the common understanding and decision-making do not subscribe to a whole-building approach, even though shallow refurbishments result in only modest energy savings. Moreover, "cream-skimming" the HVAC and other shorter term options makes future investments for remaining needed items even less appealing since the shortest term investment have already ready been done.

In fact, a DER that uses a combined bundle of ECMs with short- and long-term payback periods generally can provide a more economical opportunity. This combined "whole-building" approach can replace the HVAC system —and downsize it— since, after the DER, the building will have lower heating and cooling demands, will require no perimeter zone conditioning, and will likely offer improved comfort. This approach is most important for HVAC since other components can be specified to require high performance without compromising the economics of the systems approach (e.g., specifying code-compliant roof insulation or window criteria during component replacement will not negatively impact future HVAC upgrades). Policy must encourage decision makers to take the "whole-building" approach.

Achievements

The DER Business Guide provides replicable information to streamline the DER process. It also provides best practice approaches to improve the cost effectiveness of DER projects and to reduce the demand for public funding by integrating bankable and non-energetic LCCs into the financing scheme. The Guide describes the owner-directed business model, and a set of advanced DER EPC business models to provide a balanced set of solutions based on guaranteed and bankable energy and LCC improvements.

© The Author(s), under exclusive license to Springer Nature Switzerland AG 2019   111
R. Lohse and A. Zhivov, *Deep Energy Retrofit Guide for Public Buildings*,
SpringerBriefs in Applied Sciences and Technology,
https://doi.org/10.1007/978-3-030-14922-2_6

Next Steps to Scale up DER in Buildings

A standardization of technical and business process is needed to stimulate the demand for DER-specific research efforts, which will consequently make data from those efforts more generally available. On the policy level, there is a need for a stabile framework to increase the reliability of long-term DER investments, most importantly, to improve the accountability of non-energetic LCC in the public sector, including the building value. In many European countries, EPC is still seen as a debt-related liability so that it is excluded from access to soft loan financing instruments. Recommendations to improve such policy-making and market development activities include:

1. **DER Performance data—Research efforts needed**. To scale up DER in buildings, additional research is necessary to collect, evaluate and make available reliable data on the capital cost, investment, and LCC performance of executed DER and NZE projects.
2. **Create standard DER solutions**. To strive for economies of scale, a strong R&D effort is needed to achieve a streamlined and standardized design of construction solutions for many specific issues in highly replicable building types that reduce design, equipment, and labor costs.
3. **Recommissioning of the public accounting regulations**. LCC public accounting regulations must be adjusted to consider non-energy measures, specifically by accommodating commercial real estate standards by considering such elements as building and land value, rental incomes from users, etc.
4. **Reconsider debt accounting for EPC**. Especially in Europe, public entities consider EPC as a debt-related liability. There is a need for an initiative to standardize debt accounting for EPC, specifically by collecting performance data and changing regulations for the standard EPC.
5. **Need for dissemination of DER financing instruments and business models**. A significant improvement and change of paradigms must be put in place within the dissemination and use of EPC related business models in the public sector. In the United States, an Executive Order of the President was considered a game changer in the use of EPC in the public sector. Such kind of initiative is still missing in the European countries. In addition, regional market structures based on regional competence centers should be put in place to provide promotion and marketing activities to overcome the inertia especially in the public sector.
6. **Facilitators are needed to develop complex DER project structures**. EPC and DER facilitators act as intermediates between the financiers, who are often unable to develop DER projects. Building owners are often not familiar with the processes. Facilitators who are trained to develop complex DER project structures for building owners are needed to prepare decision-making processes, and to establish performance-related innovative business models.
7. **More confidence is needed among participants**. To direct private money into DER projects, the investor, ESCO, and building owner must share confidence. In Europe and the United States, the ICP provides a structure that allows the transparent preparation, assessment, and approval of DER projects. For smaller projects,

the SMESCo approach may be viable. It is necessary to improve and establish these confidence-creating tools quickly among the business banks, funds, and ESCo.

8. **Create innovative funding sources**. Large amounts of private money are needed to supplement scarce public funds, and ultimately to provide the investments needed to initiate a DER strategy. To achieve this end, more reliable data from accomplished DER projects are needed to provide the essential information to make precise predictions and to help foster DER projects' credibility. This can only be achieved through extensive research and evaluation projects. One of the next steps is to initiate a discussion with institutional investors and to tap the large potential available in energy cooperatives.

Appendix
Best Practice DER Business Models

A.1 DER Business Models in the U.S. Federal Sector [1]

Two different models have emerged for implementing DER projects in the U.S. federal sector: (1) through conventional EPC project, and (2) through EPC projects that are combined with comprehensive building renovations.

The U.S. Army is experimenting with the combined approach (2). The model would use two different contractors: a renovation contractor funded by appropriated funds to accomplish non-energy-related upgrades, and an ESCo, which obtains private financing to implement energy upgrades. The advantage is that the cost of envelope-related conservation measures, which are not often included in EPC projects, can be reduced by coordinating them with the activities of the renovation contractor.

For example, the ESCo's cost to replace wall cavity insulation will be lower if the renovation contract includes replacement of wallboard. However, several challenges exist: coordination of project design and construction, management of the overall project, and dispute resolution between the two contractors. While a procurement strategy exists on paper, the Army is still considering pilot sites at which to implement this approach.

The GSA has had success in reducing the energy use of its buildings (Fig. A.1) by 2012 progress had begun to stall. For this reason, GSA began a program focused on achieving deeper energy savings using the conventional ESPC process. To this end, in March 2012, GSA issued a Notice of Opportunity (NOO) for a nationwide deep energy retrofit (NDER). The NOO included a list of 30 GSA-owned buildings covering a total occupied area of 16.9 million square feet in 29 states and the U.S. Virgin Islands. Among the objectives for the project were:

- Retrofit plans that move a building toward net zero energy consumption
- Use of innovative technologies
- Use of renewable energy technologies.

© The Author(s), under exclusive license to Springer Nature Switzerland AG 2019
R. Lohse and A. Zhivov, *Deep Energy Retrofit Guide for Public Buildings*,
SpringerBriefs in Applied Sciences and Technology,
https://doi.org/10.1007/978-3-030-14922-2

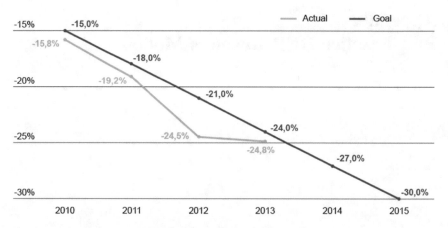

Fig. A.1 Energy intensity reduction in GSA buildings since 2010 (*Source* Oak Ridge National Laboratory [ORNL])

Evaluation of GSA's NDER Projects

GSA ultimately awarded 10 EPC Task orders with a total value of $172 million distributed among seven ESCOs. The projects covered a total of 1,365 million square meters of space in 23 buildings. It will reduce GSA's energy consumption by 108 GWh per year, resulting in a first-year guaranteed cost savings of $10.8 million, which will be used to pay pack the investment over time. A key result from the project was the average 38.2% proposed energy savings over the baselines, which is more than double the average proposed energy savings in a sample of 80 other recent federal EPC awards. Figure A.2 compares the percent of energy reduction of the GSA projects (filled circles) with the percent energy reduction of the other federal projects (open circles).

Fig. A.2 Percent energy reduction of NDER projects compared with other U.S. federal ESPC projects. *Source* ORNL

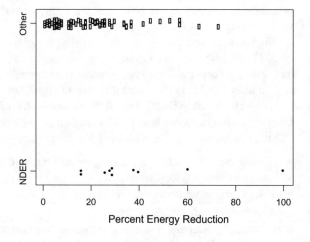

While GSA's NOO expressed a preference for innovative technologies and renewables, it is noteworthy that the majority of savings in the NDER projects were achieved using conservation measures similar to those encountered in other projects: lighting upgrades, controls retrofits, chiller and boiler replacements, etc. The assessment of the GSA NDER projects showed a couple of KPI that second the implementation of DER EPCs:

- The majority of the selected buildings should have not undergone recent energy retrofit projects and still provide the combination of "low hanging fruit" and ambitious energy and infrastructure measures.
- Framework settings for the public procurement process: So far, EPC was considered to provide energy savings between 20 and 30%. The emphasis from GSA to target DER encouraged ESCOs to propose longer-payback ECMs, and regional facility managers to accept these suggestions even when the cost effectiveness is lower than in the "business as usual" approach.
- To identify DER ECMs, the ESCOs have been requested to participate in a thorough energy audit process steered by the regional facility managers. This approach targeted an integrated design approach that considers the building, its occupants, and energy consuming equipment as a holistic system. In these concepts, the refurbishment of the building envelope was not a major factor in achieving deeper savings.
- What is not (necessarily) required to achieve deeper energy savings in EPC: The assessment of the DER projects steered by GSA led to a couple of interesting insights:

  - The level of energy savings obtained in the projects was unrelated to site energy prices.
  - The level of savings in the projects was unrelated to baseline energy intensity.
  - Large "buy downs" of appropriated funds in the form of initial payments from savings were not needed: The level of savings was unrelated to the size of the upfront payment.

A.2   DER in Belgium's Public Buildings via Advanced EPC [2]

The Belgian facilitator "Factor4" has been involved in a number of EU projects such as Transparense [3], the "European Energy Service Imitative 2020 (EESI 2020)" [4] and in IEA DSM Task 16 research work [5]. To enhance the implementation of DER in the Belgium Factor 4 developed an advanced "SmartEPC" business model. The model was implemented in 2015. The first contract period will end in autumn 2016.

Integrating Non-energetic Measures and Benefits in EPC

Investment decisions for a general refurbishment of a building are often driven by the aspiration to increase the building's functionality. Energy efficiency is merely a positive "side-effect" of a building refurbishment. However, "business as usual" EPC structures are related only to the energy savings. The integration of non-energetic measures and benefits into the scope of EPC projects intends to increase investment costs dramatically by a factor of 2 or 3. To keep the balance between investment costs and savings, "SmartEPC" inherits non-energy-related savings into the cost balance:

1. increased value of the building,
2. a higher level of indoor climate and user comfort.

"SmartEPC" provides calculation methods to make the non-energy-related savings accountable and, with regard to their bankability, gives guidance on how to assess and to verify their performance. The business model requires the fulfillment of basic project requirements (functionality of the refurbished building, safety, legal standards, etc.), but offers large decision autonomy for the ESCO in choosing the strategy to achieve these energy and non-energy-related benefits.

Concept

The decision-making criteria are like the criteria used in most of the EESI 2020 driven EPC procurements, i.e., they select the ESCO with the maximum net-cost saving. The net-cost saving is the annual guaranteed energy cost savings added to the increased value at the end of the project minus the annual remuneration of the ESCO (Fig. A.3).

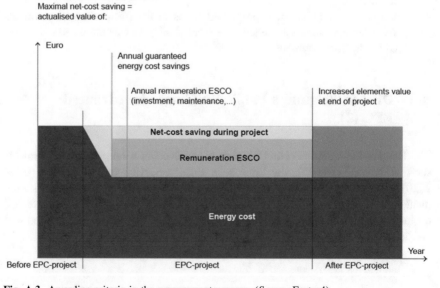

Fig. A.3 Awarding criteria in the procurement process. (*Source* Factor4)

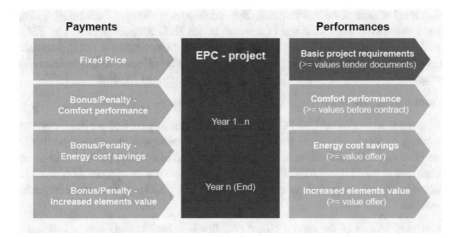

Fig. A.4 Revenues within the EPC project. (*Source* Factor4)

The "SmartEPC" accounts for the following performance criteria to determine the revenue streams and services between building owner and ESCo:

- Fixed price (payment) during the contract period related to the fulfillment of basic project requirements (maintenance of new and existing HVAC equipment, building automation) defined in procurement requirements
- Bonus-malus payments if defined comfort parameters are under- or over-performed
- Energy savings validated with fixed energy prices during the contract period related to the measured and verified energy savings during the contract period
- At the end of the contract period, a bonus-malus payment according to the increased or declined elements and building value is settled if an additional value is achieved, e.g., by appropriate maintenance, the ESCO is awarded a down payment at the end of the contract period (Fig. A.4).

Energy Performance

In the "Smart EPC" concept, ESCOs take over the energy-saving performance risk and are paid according to the measured and verified energy-saving performance—similar to "the business as usual" EPC. The money that the customer saves on energy costs (or part of them) is forwarded to the ESCo during the contract period.

Maintenance Performance

In "SmartEPC," the ESCo is technically and financially responsible for operation, maintenance, and replacement costs for the whole building including installations in and at the building such as the thermal envelope, windows, roof, HVAC, lighting, elevators, distribution grids and ducts. In exchange for taking over the risks for existing and replaced equipment, the ESCo receives a fixed price as an extended maintenance fee. "SmartEPC" provides an additional incentive for the ESCo for high level maintenance by evaluating the condition and value of the whole building [6] at the end of each year and at the end of the EPC contract period. The ESCo participates in increased values and is penalized if the condition is not appropriate. This approach provides strong incentives to conduct a sustainable maintenance program by putting in place measures with long technical lifetime.

Comfort Performance

In business as usual (BAU) EPC contracts, the indoor climate and the indoor quality is only a qualitative factor in the sense of a basic requirement that should be met by the ESCo after the refurbishment. The money value of comfort performance is not taken into account in a BAU-EPC contract, which also has an optimization potential regarding specific regulations related to high-level building comfort performance because:

- The definition of the indoor climate quality is inflexible and bears questionable parameters that may not fit to different levels of usages.
- The value of comfort aspects e.g., customer-friendliness of ESCo, is not a performance criterion.
- In the BAU-EPC approach, the building owners are mostly not involved in the design of measures that may affect the indoor comfort level. To involve the users may help to distinguish non-critical and critical comfort aspects from the buildings users' perspective.
- The M&V of the comfort performance level is expensive due to the needed metering and reporting efforts.

"Smart EPC" Allows for Accounting Comfort Performance

A DER offers the opportunity to increase the comfort performance and to create additional value for the building users and owners. For example, wall insulation and high-efficiency windows will reduce cold or hot indoor surfaces, which as a result, will allow good quality working places to be located much closer to external walls.

"Smart EPC" introduces mechanisms to increase the remuneration streams by transparently validating and monetarily quantifying the comfort performance. Comfort parameters are metered using a "Comfort meter" (www.comfortmeter.eu), which is an online questionnaire for the building users to qualify the indoor comfort conditions and to set up a **comfort score**.

The "Comfortmeter," which was developed by Factor 4 in close cooperation with two universities, reduces the investment costs compared to a large M&V program.

The ESCo guarantees a minimum comfort score; each score beyond that minimum level may increase the remuneration of the ESCo. The scale of the remuneration is calculated assuming, for instance, that an increased comfort score of +1% generates 0.2% productivity increase. This relation between comfort score and (self) reported productivity was proven via a Comfortmeter survey of 1500 employees working in 35 buildings. The Comfortmeter questionnaire polls the comfort experience of the employees via 35 comfort questions related to different comfort aspects, such as temperature, sound, and air, but also the expected effect of the comfort on their productivity. Through a statistical analysis of the 1500 survey results, the mentioned relation between comfort score and productivity could be estimated. In total, the additional remuneration from the comfort score can be up to 9 €/m²year or more.

Conclusions: Within "SmartEPC," ESCOs are more focused on higher comfort and employee satisfaction. "SmartEPC" is thus able to create additional value to contribute to the financing of cost intensive DER by: (1) monetarily quantifying comfort performance, and (2) providing a business model in which incentives for a high level maintenance program are given based on the ESCO's participation in the increased building's component value at the end of the EPC contract. Both financing contributions extend the financing scope of BAU-EPC business models significantly.

A.3 EPC Business Models in Latvia's Residential Building Sector [7]

The Housing Market in Latvia

Latvia and other east European countries from the former Soviet Union are facing serious challenges in their existing building stock. The severe housing deprivation [8] rate is more than three times higher than the EU-27 average [9]. The overcrowding rate [10] of almost 60% is the highest among the EU-27, more than three times the EU-27 average. About 60% of the Latvian people are at risk of poverty, twice as high as the EU-27 average. On the national level, Latvia's floor area per person is very limited. A further degradation would lead to a severe housing crisis. Currently, two major challenges have arisen:

- The buildings were designed in the 1960/70s to be built "cheap and fast" with an expected lifetime of ±30 years. They were not properly designed to withstand harsh weather conditions. Consequently, external parts are now corroding due to the effects of weather; panel joints are becoming crushed; balconies are crumbling; and roofs are leaking. Internal parts such as water, sewage, and ventilation networks, which were poorly designed, have become heavily corroded.
- Ninety-seven percent of Latvia's building stock is owner occupied. After the breakdown of the Soviet Union, tenants became owners of their flats. However, many people cannot afford to undertake the necessary conservation measures. Most importantly, they lack the organizational capacity to live up to their responsibilities.

RenESCo's EPC Business Model for the Housing Sector

RenESCo is a residential private ESCo and a social enterprise that finances housing modernization through energy conservation. The ESCo is driven by the challenges of the deprived and overcrowded building stock. RenESCo won the European

Energy Service Award 2011 in the category "Best Provider" for its commitment and its innovative approach.

RenESCo´s business model is based on an EPC contract, in which RenESCo takes over the whole conservation and modernization process of the apartment buildings, and also assumes responsibility for operation and maintenance for 20 years. The flat owners are obliged to pay the energy cost savings to RenESCo during the EPC period (20 years). RenESCo has the responsibility for the planning, implementation, funding, operation, maintenance, and M&V.

Roughly 60% of the funding of the projects comes from the energy cost savings financed by RenESCo, and 40% comes from the ERDF-funded national renovation program.

Sixty percent of RenESCo's share of the investment currently consists of debt-financing from a Latvian bank. The bank financing is based solely on the EPC contract. No other collateral is used. RenESCo must bring in the remaining 40% of the funding as its own equity capital.

Benefits for Residents

For the apartment owners, the momentum to engage in the RenESCo business model is only indirectly related to energy performance. Other concerns are:

- Increased indoor comfort, health, and reliability of the building, which are all part of the services provided by RenESCo.
- Conservation and modernization of apartments, which results in a 20-40% increase of the market value and directly benefits the residents.
- The refurbishment creates more comfortable and acceptable looking houses to live in. RenESCo guarantees a temperature level of 21.5 °C. (Currently, many apartments are severely under-heated).
- The flat owners incur no additional refurbishment costs.
- There is a 20-year guarantee on all construction and therefore no additional cost for maintenance during the contract period.
- After 20 years, apartment owners will have recuperated their costs from energy savings. The savings are estimated to be in a range from 50-80%.
- RenESCo offers an additional value preposition and an incentive for the apartment owners to contribute to keep the energy consumption as low as possible by offering a 25% profit share of RenESCo's net result.

Achievements

RenESCo provides a DER for the buildings:

- Within 5 years, RenESCo financed 100% of the cost and performed deep renovations of 15 typical Soviet-era apartment buildings using an EPC business model.
- The DER measures for this specific building type include the refurbishment of the complete building envelope in a thermal insulation composite system (TICS) with an average thickness of 10cm, installation of new domestic hot water and

networks, new heating network, new ventilation with heat recovery systems, and cosmetic repairs.

- The existing "natural ventilation" system creates airflow from the leaking building envelope and windows to the indoor floor area and an uncontrolled exhaust air network in the bathrooms. The new ventilation system is a mechanized ventilation system with (90%) heat recovery and a control system. In the summer, free cooling is provided. The DER reduces the building's leakage rate to 10% of the value before the refurbishment.
- Improvements are made to the heat supply, which is typically city heating. Where possible, geothermal heat pump systems with vertical probes have been installed.

An evaluation by Ekodoma and the Riga Technical University has shown that the RenESCo business model provides a high level DER. The energy-saving guarantees and the EPC contracts have proven themselves to be bankable by a local financier. RenESCo's projects clearly illustrate a successful DER that includes a wide scope of non-energy-related measures at the same or lower costs and that results in better quality than other municipal and private sector projects in Latvia.

Credit Rating of Housing Owners

A perceived barrier that must be overcome is that from financing institutions assume that low- and medium-income people will not be able to pay the bills. RenESCo's experience shows a different picture. Even during a time of economic crisis with high unemployment rates, RenESCo received 97% of payments on time, and 0% non-payment during its 6 years of operation. This can be explained by the explicit connection between apartment ownership and the heating and maintenance bills. Poor owners have a strong incentive to pay their utility bills because they will otherwise be forced to sell their renovated flats and move to flats with has similar utility costs, but less comfort.

RenESCo's Credit Rating

Apart from the doubted creditworthiness of its customers, financiers fear that the expected energy savings will not be achieved, or that they will fall after some years of performance. RenESCo's experience shows that expected and performed savings are usually within a 2–5% range of error and that they remain constant over time. Since the building stock in RenESCo's projects is more or less of the same age, energy consumption per m^2, and scope of measures, there was already a record of experience that helped mitigate performance risks (Table A.1).

Nevertheless, there is no suitable financing available. Many approved projects had to be cancelled due to lack of finance. Despite the recorded experience of reliable payments by the apartment owners and reliable predictions on the energy-saving performance, RenESCo's cost of capital is still much too high ($\sim 7\%$). Creating a forfeiting fund to buy up the RenESCo's future cash flows is considered a viable option to lower RenESCo's financing costs and to enable a quicker recapitalization, but this has not yet been put in place.

Table A.1 Real and perceived risks in Latvian housing sector (*Source* RenESCo)

Barrier/Risk type	Perceived barrier	Real or Red Herring?
Performance	Low and medium income people have to pay the bills. Many will not be able to	Proven Red Herring. Banks love to talk this up in order to increase the interest rates they can charge, or collect extra guarantees from governments. Track record of renovation finance is excellent in Eastern Europe
	Expected energy savings will not be achieved, or will drop after some years of performance	Proven Red Herring. Thousands of similar buildings Expected savings usually within 2–5% margin n of error. Savings do not decrease over time
Policy	Lack of consistent policy. Start/Stopping of programs	Major problem. It takes 1–2 years to develop projects years to develop capable organizations Stopping programs destroys projects and renovation companies
	Transection costs. Complexity of support programs. Procurement rules	Real Consumes at least 70% of RenEsco staff time, Adds 10–15% to the total project costs. Creates many unnecessary risks ait pro.eet failures Leads to silly and poor decisions.
	No financing variable. Subsidy yes, finance no. Especially problem for small private sector companies	Real Many approved projects cancelled because lack of finance 1% lower interest over 20 years = 9% investment subsidy. ESCO cost of capital is much to high (7–10%)

RENESCO's Experience in Finding Workable Programs and Finance

In the last period (2007–2013), many countries offered loan and subsidy programs that excluded third parties such as ESCo as borrower or grant receiver. The European Commission and International financial institutions went along with this, preventing ESCOS or other third parties to develop. One major challenge in the support of third private parties with public seed money or subsidies involves tight market regulations ("de-minimis") that target a market situation of "equal opportunities for all," which may not derailed by public grants. To bridge that issue, building owners should have access to public grants under the obligation that they engage an ESCo to implement the project.

The greatest barrier to implementing DER in residential buildings in most eastern European countries does not originate from the realization of energy savings. Instead it originates mainly from the lack of consistent policies from governments and financial institutions to realistically deal with the post-Soviet housing legacy. Local and international financial institutions still hesitate to give proper consideration to the example of successful projects like RenESCo's to allow, for example, EPC contracts to serve as collateral to secure financing streams. An evaluation of the practice and the development of PF structures could contribute to overcome this barrier.

A.4 Implementation of Advanced DER EPC Business Models in Dormitories in Mannheim/Germany [11]

A.4.1 Initial Situation

The building owner, the student´s union (Studierendenwerk Mannheim) is a public institution. The main aspects of the work of Studierendenwerk Mannheim are concentrated in: (1) student housing (19 apartment buildings with some 3,150 apartments), (2) catering and dining (13 canteens and cafeterias), and (3) tuition fee funding and loans—BAföG (approx. 10 million Euro in funding per year). In 2008, SW Mannheim installed an energy management structure that commissions the energy consumption of all buildings. Also a master planning process includes a 10-year refurbishment roadmap for all apartment buildings was crafted in 2008. After some experience with low performing DER projects, the building owner decided to continue the refurbishment process only if the responsibility of the energy performance were provided by the contractor. In 2015, it was agreed to set up the first German DER EPC project in the Ludwig Frank quartier.

Prepare the minimum requirements and the energy baseline in a building energy audit: The building energy audit is carried out in a staged process and prepares all data required for the EPC measure list, and for the energy and water/sewage and cost baseline. The global cost baseline is 446k€/yr, which is relatively small for a typical EPC project (Fig. A.5, Tables A.2 and A.3).

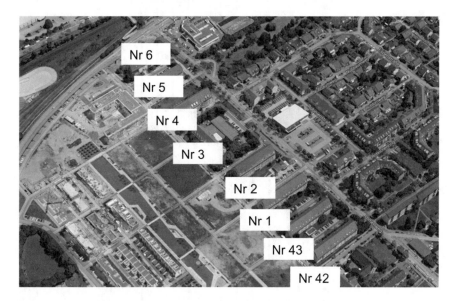

Fig. A.5 Top view on Ludwig Frank quartier

Table A.2 Ludwig Frank quartier heating baseline and benchmarks

Heating (district heating)

Building	Baseline climate adj. (kWh/a)	Baseline price (€/kWh)	Load (kW/a)	Load (€/kW)	Fixed price (€/a)	Cost baseline (€/a)	EUI_{Heat} (kWh/m²year)/ kBtu/ft²	CUI_{Heat} (€/m²year)/ (€/ft²year)
42	328,293	0.0459	42	106.21	206,83	19,736.30	95 (30)	5.7 (0.52)
43	347,907	0.0459	43	106.21	155,13	20,691.09	101 (32)	5.8 (0.52)
1	384,342	0.0459	65	106.21	155,13	24,700.08	90 (28.6)	5.6 (0.52)
2	417,097	0.0459	49	106.21	86,18	24,435.22	108 (34.2)	6.35 (0.57)
3	303,087	0.0459	144	106.21	155,13	29,361.06	131 (41.5)	12.6 (1.14)
4	308,633	0.0459	50	106.21	86,18	19,562.93	78 (24.7)	4.97 (0.45)
5	237,184	0.0459	38	106.21	88,16	15,010.89	81 (25.6)	5.1 (0.46)
6	236,777	0.0459	38	106.21	88,13	14,992.17	88 (27.9)	5.8 (0.52)

Table A.3 Ludwig Frank Quartier buildings

No.		Year of construction, inhabitants	Floor space	Recent refurbishment
42, 43		1960 97 in.	2.666 m^2 (28.697 ft^2)	1994 Windows partly refurbished, attic floor insulated, new roof tiles
1, 2		1933 139 inhabitants	3.691 m^2 (39.730 ft^2)	1998 Windows partially refurbished, 5 cm mineral wool on attic floor
3		1933 44 inhabitants, Restaurant in ground floor	2.790 m^2 (30.032 ft^2)	1998 Windows partially refurbished, 5 cm mineral wool on attic floor
4		1933 110 inhabitants	3.200 m^2 (34.444 ft^2)	1998 Windows partially refurbished, 10 cm mineral wool on attic floor, basement ceiling 5 cm, roof tiles, entrance doors
5, 6		1960 88 inhabitants	2.257 m^2 (24.294 ft^2)	1998 Windows partially refurbished, 10 cm mineral wool on attic floor, basement ceiling 5 cm, roof tiles, entrance doors

A.4.2 De-Risking Approaches

A DER EPC in which an ESCo takes responsibility for the energy performance of a thermal retrofit has not been carried out before. The goal of the project is to carry out the DER concept in combination with a performance guarantee in which the remuneration system is related to the verified performance of the implemented project measures. A DER EPC that includes the holistic refurbishment of a building envelope has not yet been carried out in Germany. During its working phase, the German IEA Annex 61 ST B working group organized three workshops with the ESCO association VfW (German Association of Heating Suppliers, Chapter EPC; www.vfw.de) and four interested ESCOs that identified a number of DRMs:

- **QA of the modeling process (DRM1)**: The evaluation of 12 German and Austrian DER projects [12] indicated that one major reason for the failure of modeling results was that they lacked back-calibration. It was agreed that KEA is providing a PHPP role model in the software PHPP, which depicts the pre-refurbishment status of the building, and which also provided one potential technical modeling solution to achieve −55% of energy savings. This database was provided for the ESCOs in the tendering and procurement data basis.
- **De-Risking of the DER project by pool building (DRM2)**: With regard to the performance-related remuneration, the major concern of the ESCOs was the risks coming from a significant failure of the calculated savings combined with high investments. If the DER building does not perform appropriately, the risks can be mitigated by the results of seven other buildings with only standard ECMs put in place. Still under discussion is an idea in which major parts of the savings are fixed after their success has been proven for 3–5 years.
- **Monitoring and Verification process-tolerance (DRM3)**: In early projects, some ESCOs received contract awards by promising high savings without actually ever intending to achieve these performance values. Many standard European EPC contracts now penalize unfulfilled performance guarantees. These penalties have been excluded from the DER EPC contract for the first 3 years of the contract period to allow the project a "learning curve" for the ESCOs. The remaining risk of receiving only the verified energy savings is not touched by DRM3.
- **Bidding cost reduction (DRM4)**: One part of the risks involves the expenditures of the ESCOs during the bidding process. These costs have been minimized by providing a "basic solution" with a functional description of all necessary details of the design of the external walls, the windows (size, format and position in the wall). Also, one cost-optimized technical specification was provided that fulfilled the minimum requirements. This approach avoids the cost created when ESCOs spend engineering costs to design minimum requirements.
- **Investment cost and QA of planning (DRM5)**: In an EPC, the ESCOs are also responsible for the total investment costs of the measures. The tendering of materials provides a functional description of the insulation measures. To reduce the risks for the DER measures, the ESCOs are required to transfer the

functional description of the thermal envelope into a more detailed specification, to collect bids from subcontractors, and to refer to this in the tendering process. The reliability of the specification and the subcontractor prices are reviewed by a small team of experts (building owner, three experienced architects). In the first DER EPCs, this mandatory quality check helped to improve the reliability of the planning and investment calculation of the DER and to avoid cost increases.

- **Collect "avoided maintenance costs" from refurbishment measure plan (RMP)**: The RMP considered the technical condition and obvious need for action of the walls, roofs, windows, HVAC installations with regard to their technical lifetime. From national reference figures [13], the refurbishment and maintenance costs were calculated over a time line of 5 years. For each building, these investment costs were cumulated and, with an interest rate of 4% over time period of 20 years (equivalent to their average expected technical lifetime), transferred into annuities of total 44 k€/year. These annuities, which reflect the need for action to keep the buildings functional, are considered as avoided building maintenance costs; ESCOs can take this quantity into account for ECMs that replace or refurbish these components (see DRM6).

- **Limitation of maintenance and replacement cost risks (DRM7)**: In most standard EPC contracts, the ESCo takes the responsibility for the availability, functionality, and energy efficiency of the measures the ESCo has implemented. In many cases, the ESCo services also include a troubleshooting service. All these responsibilities exceed the "normal" guarantee provided by manufacturers. The incorporated risks are well known and available as empirical data that reflects each ESCo's practical knowledge of HVAC measures. This project will implement a combination of HVAC and DER measures. The individual measures have a technical life time expectation of 8 years (for a building automation device, speed controlled hot water pumps) and 30 years (for windows) if German standards [14] are considered. DER EPC will have extended contract periods (in comparison with HVAC EPCs) that will automatically lead to increased and unforeseeable risks. In the first DER EPC, these risks are limited to one life cycle for larger HVAC parts, two life cycles for smaller HVAC devices, and to certain responsibilities in the DER investment parts. (For the maintenance of the windows and the thermal envelope only a limited cost capacity of 0.3 €/m^{2}year must be provided.) In case this is not sufficient, the building owner will pay the difference needed to maintain the functionality.

- **Decision-making criteria for the tendering process (DRM8)**: this approach refers to the integration of non-energy-related criteria [15] into the decision-making process (Table A.4).

- **Technical specification**: The technical specification should provide all necessary information that helps to limit the effort for the ESCOs to calculate. However, the specification should invite the ESCo to provide their own ideas in the bidding process. Hence a functional specification is provided with a description of the boundaries and interfaces, and technical functionality. The design, color, and shape of the external walls; the window partition; color and measures of the frame are all described in all details to avoid any

Table A.4 Award Criteria of business as usual EPC and DER EPC Mannheim

	German EPC (business as usual)	DER EPC
Award criteria for EPC tendering	**(1)** **N**et **P**resent **V**alue of savings in total and remaining with administration 70–80%	(1) Net present value of savings in total and NPV of the partition of the savings remaining with building owner 50%
	(2) Contract period 10–20%	(2) Sustainable measures and Concept 40%
	(3) Carbon Footprint 10–20%	(3) Carbon Footprint 10%
Additional terms	–	Avoided maintenance costs for the replacement of existing installations are part of the saving

misinterpretation. In addition, the functional specification also provides the minimum requirements for the building U-values with reference to the KfW standards KfW 100 (which equates to the energetic quality of a new building) [16] and the definition of minimum HVAC measures.

- **Transparent tendering process**: To accommodate highly complex EPC projects (e.g., integration of biomass and infrastructure measures), this DER EPC tendering process is to be conducted in three stages:

  - Selection of three ESCOs with primary experience in ECMs at the thermal building envelope and with experience in the use of modeling tools at least on a monthly basis.
  - Tendering, negotiation, and EPC contract award: the selected ESCOs receive the contract and process documents including the baseline and the functional specification, and create their technical concept and commercial bid. Both will be presented in two negotiations, decision-making will then be prepared and one ESCo will be awarded. This is happening at this point of time (June 1, 2016).
  - Detailed planning phase: the awarded ESCo together with SW Mannheim will prepare a detailed technical plan. External experts with expertise in DER and building physics will provide the QA of the technical plan. After agreeing on the detailed planning, the second implementation phase will begin in September 2016.

- **Least-cost planning—Overview of results**: The bundling of ECMs and energy supply measures (ESMs) can increase the cost-effectiveness of a DER EPC significantly. That cost-effectiveness will consequently reduce the investment and performance risks for the ESCOs. In the Ludwig Frank quartier, the combination of HVAC measures and PV in seven buildings, along with the CHP and a DER in one building can provide a dynamic payback period of global investment costs in 17.4 years without any seed money. Table A.5 lists the cumulated impacts of these measures on the payback (Fig. A.6).

Table A.5 Cost optimization of the DER EPC project at Ludwig Frank quartier

	Investment	Energy savings	E-cost savings K€/Yr (€/m²year)	Cumulated static payback (years)
B 42 DER envelope, ventilation, HVAC, lighting	982 T€ (263 €/m²)	57% (54% heating, 3% power)	11.5 k€ (3 €/m²year)	91
B2 HVAC, lighting, PV	118 T€ (36 €/m²)	43% (10% heating, 33% power)	4.5 k€ (1.5 €/m²year)	75
B3 HVAC, lighting, PV	118 T€ (35 €/m²)	26% (10% heating, 16% power)	6.1 k€ (2.1 €/m²year)	61
B4 HVAC, lighting, PV	115 T€ (35 €/m²)	31% (10% heating, 21% power)	5.1 k€ (1.8 €/m²year)	55
B5 HVAC, lighting, PV	105 T€ (33 €/m²)	46% (10% heating, 36% power)	15.1 k€ (5.2 €/m²year)	40
B6 HVAC, lighting, PV	119 T€ (38 €/m²)	41% (15% heating, 26% power)	11.2 k€ (3.8 €/m²year)	35
B7 HVAC, lighting, PV	120 T€ (38 €/m²)	53% (15% heating, 38% power)	12.7 k€ (4.0 €/m²year)	31
B8 HVAC, lighting, PV	115 T€ (35 €/m²)	56% (15% heating, 41% power)	6.6 k€ (2.2 €/m²year)	31
Buildings 1-8 supply solution CHP, Gas peak boiler,	+749k€ (overall 28 €/m²)		113 k€/year (overall: 4.2 €/m²)	20
Total	2.541 k€ (93.1 €/m²)		185.8 k€ (7.2 €/m²year)	
Partition of avoided M&R to achieve 17 years			44 T€/year (1.6 €/m²)	13.8 years (static payback)

A.4.3   *Conclusion of the Advanced EPC Case Study Mannheim*

The facilitation of the first German DER EPC has provided significant progress in terms of the development of EPC as an instrument dedicated to take advantage of "low hanging fruits" using a sustainable vehicle that can help implement European legislation (EBPD) and national implementation strategies. The following conclusions may be drawn from this DER EPC case study:

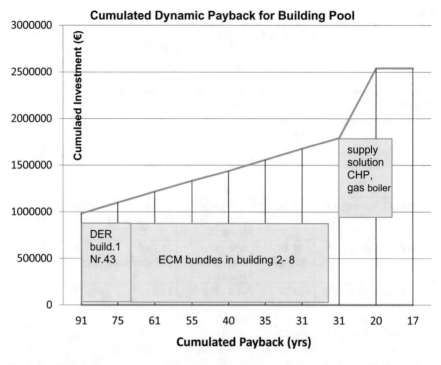

Fig. A.6 Cumulated investment over cumulated payback period for the DER EPC project in Ludwig Frank quartier

1. A DER EPC can be feasible and can attract ESCOs if certain risks are made transparent and are distributed between the building owner and the ESCOs. In this case study, the major de-risking steps were:

   a. maintenance costs for equipment with a technical life time < than the contract period
   b. simplification of M&V process
   c. the planning and design of minimum and architectural details in the preparation of the tendering
   d. transparent tendering and decision-making process
   e. the setup of a building pool with combined DER and HVAC measures.

2. Cost effectiveness can be achieved by integrating non-energetic life-cycle costs into the performance scheme and by combining DER with renewable energy sources for self-sustainable use in the buildings.

References

1. Development of new business models to integrate deep refurbishment; John Shonder, Oak Ridge National Research Laboratory
2. Comfort Meter project report, Johan Coolen, Factor 4, Antwerp, 2016
3. www.transparense.eu/be
4. www.eesi2020.eu/be
5. www.ieadsm.org/
6. NEN 2767 Dutch standards
7. Lohse et al. Investing in Energy Efficiency in Buildings, Brussels, 2015 based on. Eric Berman, RenESCO Project Report, Riga, Latvia (2013)
8. Lohse et al. Investing in Energy Efficiency in Buildings, Brussels, 2015 based on Eric Berman RenESCO, Riga, Latvia, 2013. Severe housing deprivation refers to the percentage of the population in a dwelling which is considered as overcrowded and exposed to at least one of the following three housing deprivation measures: A leaking roof or damp walls, floors, foundations or rot in window frames or floor, lack of a bath, shower, or indoor toilet
9. Lohse et al. Investing in Energy Efficiency in Buildings, Brussels, 2015 based on .RenEsco calculations based on EU-SILC 2009 – revision 1 of August 2011
10. Lohse et al. Investing in Energy Efficiency in Buildings, Brussels, 2015 based on RenEsco calculations based on EU-SILC 2009 – revision 1 of August 2011, A person is considered as living in an overcrowded dwelling if the household does not have a minimum number of rooms per person
11. ST C Pilot Case Study Report, Rüdiger Lohse, Martina Riel (2016)
12. EDLIG, Arbeitspaket A-interim report of the assessment of 12 DER projects in Germany and Austria (draft), KEA, Karlsruhe (under review) (2016)
13. BKI, Baukostenindex Deutschland, Band 1–6, Berlin, 2014- this investment cost index is collected from more than 10,000 different project related data sources
14. VDI 2067, B1 available at https://www.beuth.de/de/technische-regel/vdi-2067-blatt-1/151420393
15. Capturing the multiple benefits of Energy Efficiency, Chapter 4 IEA, 2015, Paris; www.iea.org/publications
16. Kreditanstalt für Wiederaufbau, KfW, Frankfurt, Germany

Bibliography

1. ASHRAE. Procedures for Commercial Building Energy Audits, Atlanta (2011)
2. http://www.buildup.eu/sites/default/files/content/BR10_ENGLISH.pdf
3. Average performance value collected from Danish Building Institute
4. http://www.bbsr-energieeinsparung.de/EnEVPortal/DE/Archiv/WaermeschutzV/
   WaermeschutzV1977/Download/WaermeschutzV77.pdf